AF375215

FROM RAW TO REFINED: THE ART AND SCIENCE OF DATA ENGINEERING

Srinivas Murri

Made with ❤ on the Notion Press Platform
www.notionpress.com

PREFACE

Data has found its way as the cornerstone to innovation and decision making, and in today's era of digital transformation. Data engineering is a field that helps shape how organizations use data for competitive advantage. This book, ***From Raw to Refined: The Art and Science of Data Engineering***, is a complete resource for those wanting to understand the detailed processes of converting raw data into tangible knowledge.

The book covers foundational principles to advanced topics, and everything in between, about data engineering. This work is structured around data acquisition, storage and processing, and data quality, security and visualization. The chapters emphasize theory and practical applications, and are crafted to address the dynamic challenges and opportunities of the data driven world.

It's more than a book, it is a culmination of knowledge that celebrates the growing art and science of data engineering. In exchange, I hope this book will provide readers with the means to help them succeed in this field.

DEDICATION

To the remarkable individuals who have shaped my journey—my partners, stakeholders, leaders, mentors, and colleagues—this book is a reflection of the lessons I've learned from all of you and the power of what we can achieve together.

From the early challenges to the exciting moments of breakthrough, your support, guidance, and belief in me have been the foundation of my growth. Each one of you has helped me navigate this ever-evolving field and pushed me to be better every step of the way.

I dedicate this book to you, with heartfelt thanks, for your unwavering trust, inspiration, and for making this journey possible.

ACKNOWLEDGEMENT

It has been satisfying and educational to write this book because many people along the way encouraged and supported me. I would first like to thank all my colleagues and peer who contributed, as a mentor, to the content and direction of this work.

I would like to take a special opportunity to thank the editorial and publishing teams for their dedicated and unwavering commitment to excellence. Your detail was too meticulous to bring this manuscript to life.

I would like to thank my family and friends who have been, my pillar of strength with your patience, understanding and motivation through this process. You've believed in my vision every step of the way, so I have been kept inspired.

Finally, I want to thank the readers and practitioners of data engineering. This transformative field is your quest for knowledge and innovation leading drive to progress. This book is for all those who are trying to make sense of data.

CONTENTS

Preface   *3*

Dedication   *4*

Acknowledgement   *5*

1. Introduction to Data Engineering   **11**

   1.1 The Evolution of Data Engineering   11

   1.2 The Role of Data Engineers in Modern Enterprises   16

   1.3 Key Concepts and Terminologies   19

   1.4 Data Engineering Lifecycle   26

   1.5 Essential Skills and Tools for Data Engineers   30

   1.6 Data Engineering vs. Data Science   39

   1.7 Challenges and Opportunities in Data Engineering   42

Multiple Choice Questions (MCQs)   53

2. Data Acquisition and Storage   **56**

   2.1 Understanding Data Sources   56

   2.2 Techniques for Data Extraction   69

      2.2.1 Data Extraction Tools in 2024   77

   2.3 Data Ingestion Frameworks   80

2.4 Relational Databases vs. NoSQL Databases 85

2.5 Cloud Storage Options 92

Multiple Choice Questions (MCQs) 109

3. Data Processing and Transformation 112

3.1 Data Cleaning and Preprocessing 112

3.2 ETL (Extract, Transform, Load) Processes 121

3.3 Real-time Data Processing 129

Multiple Choice Questions (MCQs) 139

4. Data Integration and Aggregation 142

4.1 Data Validation and Error Handling 142

4.2 Overview 147

4.3 Data Integration Techniques 148

4.4 Data Aggregation Strategies 158

4.5 Ensuring Data Consistency 161

Multiple Choice Questions (MCQs) 165

5. Data Quality and Visualization 169

5.1 Defining Data Quality Metrics 169

5.2 Data Validation Methods 176

5.3 Implementing Data Quality Frameworks 180

5.4 Principles of Effective Data Visualization 187

5.5 Tools for Data Visualization 201

5.6 Crafting Meaningful Reports 215

Multiple Choice Questions (MCQs) 218

6. Data Security and Privacy   221

    6.1 Data Encryption Techniques   221

    6.2 Access Control Mechanisms   230

    6.3 Compliance with Data Privacy Regulations   239

Multiple Choice Questions (MCQs)   242

7. Advanced Topics in Data Engineering   246

    7.1 Machine Learning Integration   246

    7.2 Big Data Technologies   252

    7.3 Data Engineering in the Cloud   260

    7.4 The Future of Data Engineering   272

Multiple Choice Questions (MCQs)   277

Bibliography   *281*

About the Author   *285*

Chapter 01

INTRODUCTION TO DATA ENGINEERING

1.1 The Evolution of Data Engineering

Data engineering has seen significant change and is now essential for data-centric decision-making. Data engineering has come a long way from its beginnings, when storing and retrieving data was a laborious and antiquated process, to its current state, where it utilises cutting-edge technology to process and analyses vast amounts of data. The voyage is a reflection of how data systems are becoming more sophisticated and advanced technologically. Data engineering has evolved from its original definition as database administration, which focused on arranging and storing data. Data integration, warehousing, and real-time processing were formerly beyond the scope of the discipline, but these areas were included as data volume, diversity, and velocity proliferated. Data engineering has emerged as a key enabler of big data analytics, which in turn drives company strategy and innovation (Reis & Housley, 2022).

Evolution of Data Engineering – The Past

Data engineering has come a long way, reflecting both the development of technology and the changing demands of organisations. Data perception and utilisation were profoundly affected by the introduction of new technologies at each stage.

- **1950s-1960s: The Beginnings with File-Based Systems:** Data engineering started out small and focused on file-based systems. Physical media including paper records, punched cards, and magnetic tapes were used for data storage. Data management was labor-intensive in the beginning due to the need for manual input and retrieval in these systems. Data was often duplicated or stored in isolated silos throughout this period, which contributed to a general lack of efficiency and standardisation.

- **1970s: Advent of Relational Databases:** The advent of relational databases in the 1970s was a watershed moment that altered the data engineering scene forever. In his relational approach, Edgar Codd suggested using tables (relations) to store information and using primary and foreign keys to keep track of links between tables. This breakthrough paved the way for the creation of RDBMSs like Oracle, which facilitate the storing and retrieval of data in a more organised and effective manner. The widespread use of Structured Query Language (SQL) as a strong and versatile tool for data processing significantly altered the landscape of data access.

- **1980s: Growth of Personal Computing and Networking:** Data processing became more accessible and decentralized with the advent of personal computers in the 1980s. Data management became much easier for companies of all sizes with the advent of personal computers loaded with database software. The expansion of LANs also made it easier to link and share data across computers, which in turn led to more integrated and collaborative data management strategies. Additionally, client-server architectures began to take shape around this time, allowing for remote access and management of databases and therefore increasing operational efficiency and flexibility.

- **1990s: Onset of the Internet and Data Warehousing:** The amount of data produced and consumed increased dramatically in the 1990s, when internet use became widespread. To manage the ever-increasing data load, organisations need increasingly advanced technologies, which prompted the creation of data warehouses. The goal in creating these centralised repositories was to provide a single, all-inclusive platform for analysis and reporting by collecting and storing data from many sources. Given this environment, ETL became crucial as it allowed for the methodical loading of data into the warehouse after extraction, transformation, and processing from several operational systems. Business intelligence and decision-making were bolstered by data warehousing's sophisticated analytical capabilities.

- **2000s: Big Data and Advanced Analytics:** Big data, defined by data volumes, types, and rates of velocity never seen before, entered the picture in the new millennium. To handle the volume and variety of data, new technologies like NoSQL databases and Hadoop, a framework for distributed processing, evolved. Data engineering solutions that are both scalable and adaptable, able to handle both structured and unstructured data, were built on top of these technologies. Data processing and analytics in real-time have taken centre stage because companies need quick insights to make smart choices. Data engineering has come a long way since its inception; incorporating data mining, advanced analytics, and ML into the field allowed for more sophisticated data-driven strategies and predictive modelling.

Evolution of Data Engineering – The Present

Modern data engineering is much more complicated and sophisticated than its early days could have ever dreamed. It currently forms the basis of data-driven decision-making that is essential to the digital economy, supporting innovation, strategic planning, and company management.

- **Integration of Cloud Technologies:** Modern data engineering is defined by the extensive use of cloud computing. Storage, processing, and analysis of data may be done efficiently, on a scalable, and cost-effective basis using cloud platforms such as AWS, Google Cloud, and Azure. This allows data engineers to handle large datasets and complicated processing jobs without requiring a lot of resources on-premise. By removing physical distance from data access and use, the cloud also makes it simpler for worldwide teams to share and collaborate on data.

- **Data Lakes and Real-Time Processing:** Data lakes are a game-changer in the world of data management. In contrast to data warehouses, which meticulously organize structured data, data lakes may store semi-structured and unstructured data, including social media material, logs, and information from the internet of things (IoT), without necessitating a preset schema. This makes it possible for businesses to gather and use more data for analysis and insights. Technologies like Apache Kafka and Spark have made it possible to stream and analyse data as it is created, giving organisations instant insights and the capacity to react quickly to changes in the market. As a result, real-time data processing has also gained significant attention.

- **Advanced Analytics and Machine Learning Integration:** Data analysis and forecasting capabilities are improved by data engineering's growing integration of machine learning and advanced analytics. In order to facilitate activities like trend analysis, operational optimisation, and consumer behaviour prediction, data engineers collaborate closely with data scientists to design and build data pipelines that feed into ML algorithms. As a result of this cooperation, positions that bridge the gap among data science and data engineering have emerged, such as ML engineers.

- **Automation and Data Governance:** Significant progress has been made in data engineering automation with the creation of tools and platforms that automate several parts of building, monitoring, and maintaining data pipelines. By reducing the need for human intervention, this improves precision and efficiency. Consistent with this, data security and privacy, as well as compliance with regulations, are receiving more and more attention. Data lineage tracing, access restrictions, and encryption are becoming essential components of data management systems that data engineers must include to guarantee compliance with ethical and legal requirements.

- **The Democratization of Data:** Data democratization is a hallmark of the current data engineering era. Data is now more accessible than ever before, thanks to self-service analytics tools and platforms. Even individuals without extensive technical knowledge may execute complicated data analysis. This change promotes a data-literate culture by giving more people in the organisation the ability to use data when making decisions.

Evolution of Data Engineering – The Future

Data engineering is set to undergo further changes in the future as a result of new technologies, more complicated data, and shifting business requirements. Data engineering is going to be influenced by a few major themes in the future:

- **Artificial Intelligence and Automation:** Machine learning (ML) and artificial intelligence (AI) will become increasingly more important in data engineering in the future. Automation driven by AI will optimize data pipelines, cutting down on mistakes and human labour. Data engineers will be able to concentrate on more strategic projects like designing data architectures and decision-support systems as a result. The efficiency and dependability of data operations will be improved by proactive prediction and resolution of data problems by advanced AI algorithms.

- **Edge Computing and the Internet of Things (IoT):** Data engineering will change as a result of the growth of IoT and edge computing, which will concentrate on real-time data processing at network edges. The amount of data created at the edge will increase dramatically as devices become smarter and more connected. More dispersed and localized data processing skills are required to enable real-time decision-making and activities, which means data engineering must adapt to efficiently handle and analyse this data.

- **Quantum Computing:** Quantum computing has the potential to completely transform the way data is processed, even if it is still in its early stages. Data engineers may soon be able to analyse massive datasets in a matter of seconds, due to quantum computing's ability to analyse complicated data sets at an exponentially quicker rate than conventional computers. This has the potential to revolutionise data-driven decision-making by enabling lightning-fast insights and answers.

- **Data Privacy and Sovereignty:** Future data engineering methods should put an emphasis on data security and privacy in light of the ever-increasing importance of these issues. GDPR and the California Consumer Privacy Act (CCPA) are only the tip of the iceberg. Global data sovereignty and privacy standards are expected to be increasingly severe, which means that data engineering processes will need to include more complex data governance and compliance procedures.

- **Collaborative Data Ecosystems:** Collaborative and open data ecosystems, in which data sharing is standardised and frictionless across organisational boundaries, are anticipated to be increasingly common in data engineering in the future. New data-driven goods and services will be developed more quickly as a result of this encouraging innovation. The proliferation of data markets and exchanges will pave the way for more efficient and ethical cross-sector data sharing.

- **Integration of Multi-cloud and Hybrid Environments:** Hybrid and multi-cloud cloud systems will dominate as organisations strive to optimize their data strategy and avoid being locked into one provider. No matter where the data sits, data engineers must manage it flawlessly across all of these contexts to guarantee consistency, security, and accessibility.

1.2 The Role of Data Engineers in Modern Enterprises

The ever-growing world of technology and data-driven decision-making has brought about significant changes to the work of data engineers. From basic data administration in the past to today's intricate big data settings, data engineers are increasingly essential to how firms utilise data to get strategic insights. This blog will explore the ever-changing data engineering sector, highlighting important features that show how data engineers are being used in modern companies.

The increasing relationship between data science, business intelligence, data engineering, and data is another important factor. In today's enterprises, these fields are not seen as separate entities, but as essential parts of a cohesive data ecosystem. Data scientists, analysts, and data engineers collaborate to make sure that data is effectively stored and converted into insightful knowledge. Data scientists' analytical skills are enhanced by data engineers' knowledge of data infrastructure management, leading to better decision-making. This partnership fosters a win-win partnership.

The scalability and flexibility of cloud solutions are also crucial for data engineers. A critical component of contemporary data engineering is the capacity to adapt resources on the fly to fluctuating data volumes and processing demands. Platforms for the Cloud

The Data Explosion

The proliferation of IoT devices, the pervasiveness of social media, and the growth of online transactions have all contributed to a dramatic increase in the amount of data produced in the last several years. The quantity, velocity, and variety of data have reached unprecedented heights as a result of this explosion. Companies must adopt more sophisticated methods to manage and derive value from these enormous datasets as conventional data management systems can't keep up with the sheer amount of data being produced. Every day, petabytes of data are created, which is just mind-blowing. The data ecosystem is constantly growing due to the proliferation of real-time information generated by the Internet of Things, updates from social media platforms, and online transactions. Businesses must acquire adaptable and trustworthy storage solutions if they are to manage this enormous volume.

The rapid and ever-changing flow of data is fueled by online transactions, IoT devices, and social media feeds, leading to an equally astounding rate of data production. Fast, fast, and data-processing systems with the ability to analyse and make decisions in real-time are essential for keeping up with the current rate of data generation.

- **Volume:** Scalable and reliable storage solutions are required due to the massive amount of data generated by many sources, including user interactions and sensor inputs.

- **Velocity:** A system that can quickly process real-time data streams is essential for getting insights quickly and making decisions quickly.

- **Variety:** Analytics solutions need to be versatile enough to handle different kinds of data due to the wide variety of data formats and sources.

Complexity of Data Ecosystems

Data management in the modern day is incredibly hard due to exponential expansion and a broad variety of data sets. Once reliable, traditional data management technologies and processes are now overwhelmed by the complexity of today's data ecosystems. One of the numerous reasons that adds to this difficulty is the massive amount of data. Scalable solutions that can manage enormous datasets are required to deal with the daily deluge of data, which includes both structured databases and unstructured multimedia information. There is already a lot of difficulty without adding in the wide range of data sources and formats.

The IoT, social media, and cloud-based platforms are just a few of the modern sources that produce Data Engineers, and they all have their own distinct format and organisation. In order to stay up with this dynamic environment, businesses need tools that can easily integrate, process, and analyse data from any format or source. Thus, companies are on the lookout for robust and flexible solutions to successfully traverse the many obstacles offered by the dynamic nature of modern data ecosystems.

Figure 1.1: The Evolving Role of Data Engineers in Modern Businesses

Source: *- (Alagar, 2023)*

1.3 Key Concepts and Terminologies

The field of data engineering comprises a wide range of ideas and terminology that are very important for the efficient management and processing of data. The following is a list of important phrases that every data engineer ought to be aware with:

1. **Data Pipeline**

    Data pipeline, at its most basic, is a representation of the movement or flow of data inside your company. It details the processes involved in gathering raw data from several sources, cleaning it up, and then sending it to an analytical destination like a data lake or warehouse.

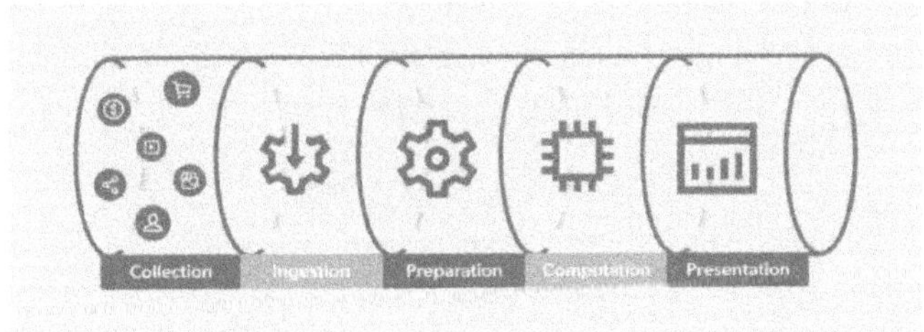

Figure 1.2: Data Pipeline

Source: *- (Nadav Roiter, 2024)*

2. Database, Schema, and Table

Any vast quantity of structured data can be stored and accessed through a database. The data can be efficiently retrieved, updated, and altered using this method. At least one schema is present in every database. The conceptual framework or "blueprint" of a database is called its schema. Tables, relationships, restrictions, and the overall structure of the database are all defined in it. Additionally, it offers a structure for specifying the data storage method, the relationships between entities, and the regulations that control the database's operations. The data is stored in a tabular format with rows and columns in a database object called a table. A distinct piece of information is represented by each column, and each row contains one item of that information. When creating a table, it is common practice to use a schema that details the table's name, column names and data types, and any data enforcing constraints or rules.

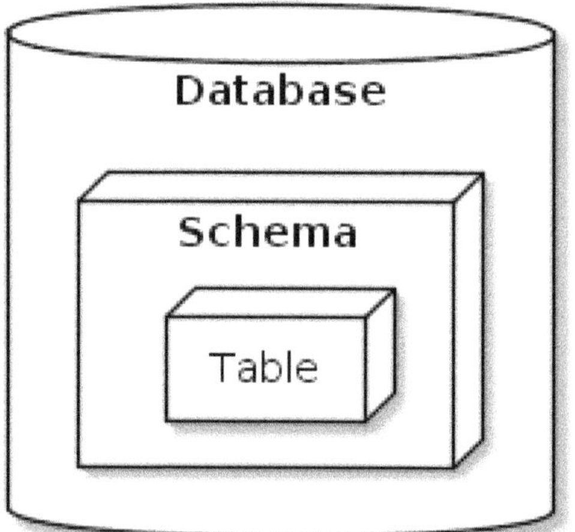

Figure 1.3: Database, Schema and Table
Source: - *(Justin Ellingwood, 2023)*

3.   **ETL and ELT**

ETL (Extract, Transform, Load) and **ELT (Extract, Load, Transform)** are two common approaches for data integration and processing.

In **ETL** approach, data is fed into a target system, such a data warehouse, after first being retrieved from different source systems and then cleaned or converted in accordance with pre-established business rules and data models.

In **ELT,** without performing any additional changes, the retrieved data is instantly put into the destination system. The processing capacity and capacities of the target system will undergo the actual metamorphosis.

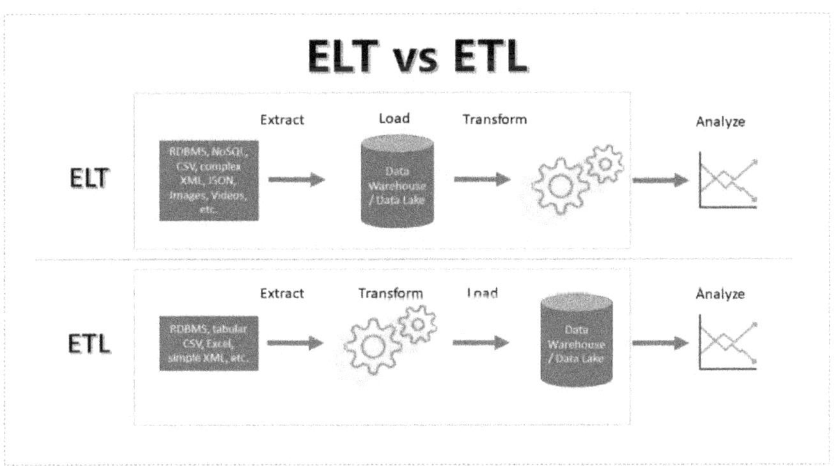

Figure 1.4: ETL and ELT

Source: - *(Edwin Sanchez, 2022)*

4.   **Data Lake, Data Warehouse, and Data Mart**

Data lakes are repositories created to hold unstructured, semi-structured, and structured data in their original format. Without enforcing a particular design or structure, a data lake's main function is to store vast amounts of data in their original format.

A data warehouse is a type of repository intended for the storage of structured data. A consistent structure is achieved for analysis and reporting by organising the data according to a predetermined template. Data warehouses are built to handle sophisticated analytical operations and are optimised for query performance. Data marts are a streamlined version of data warehouses that are tailored to meet the reporting and analytical requirements of particular business functions.

5. Batch and Stream Processing

Two distinct methods exist for processing data: batch processing and stream processing. Data is gathered and processed in a scheduled manner in batch processing. Minute, hour, daily, weekly, or monthly processing options are available. Its primary use case is in processing data offline without needing real-time or instantaneous analysis. Stream processing, in contrast, is concerned with processing data in real-time. Processing and analysing data in a linear and organised fashion is a key component of this process. Applications that require immediate attention can benefit from stream processing systems, which allow for analysis and reaction to data streams in real-time.

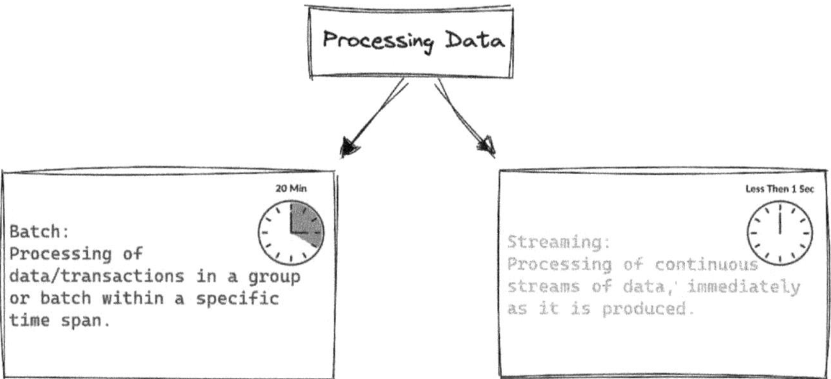

Figure 1.5: Batch and Stream Processing

Source: *- (Chandra, 2023)*

6. Data Quality

The term "data quality" describes how well data satisfies the criteria and needs of its intended application. It covers a lot of ground, including being precise, comprehensive, consistent, valid, timely, and original.

Figure 1.6: Data Quality Dimensions

Source: *- (AnalyticsPartners, 2023)*

7. Data Modeling

Data modelling is planning the structure of data storage, retrieval, and analysis in order to maximize efficiency. The three main categories of data models are physical, logical, and conceptual. As of this writing, dimensional modelling remains the gold standard for representing raw data in models.

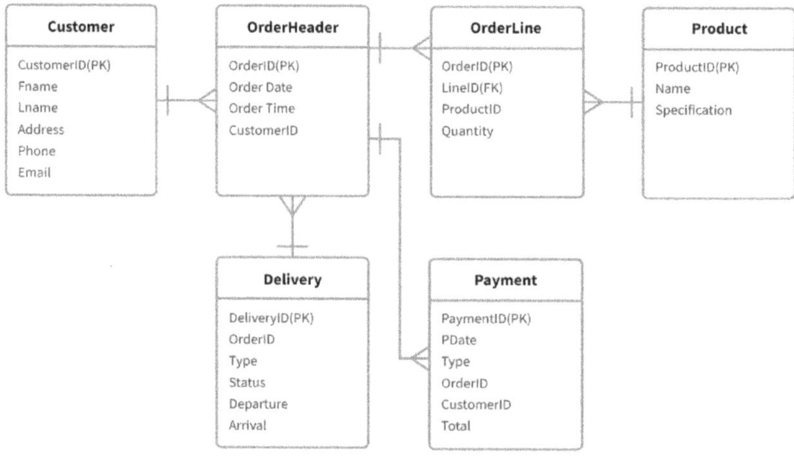

Figure 1.7: Data modeling

Source: - *(Qlik, 2023)*

8. Data Orchestration

The term "data orchestration" describes the method by which a company's different systems and applications are able to transfer, alter, and integrate data. It guarantees uniform and smooth data flow amongst all parties involved, including data consumers, storage systems, and data providers. The Apache Airflow, Prefect, and Mage data orchestration frameworks are among the most widely used.

9. Data Lineage

Data lineage traces the history of data, from its inception to its many changes and final resting places. It records and follows the data as it travels from its origin systems to its destination systems. The process of data use, transformation, and integration inside an organisation can be better grasped with its help.

Figure 1.8: Data Lineage

Source: - *(Qlik, 2023)*

10. Git

A version control system called Git enables you to work together on the same codebase with your coworkers and monitor the modifications you make to the code over time. Git also lets you give and receive feedback on each other's work in real time. You can simply use Git to roll back to the prior version of the code and restore functionality if your new code modification is disrupting your data pipeline.

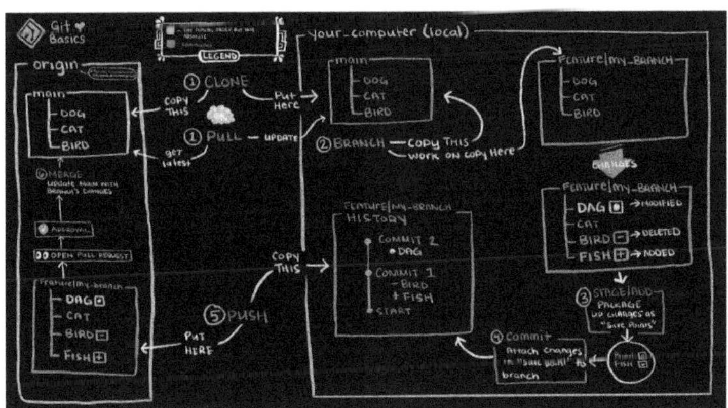

Figure 1.9: Git

Source: - *(Getdbt, 2023)*

1.4 Data Engineering Lifecycle

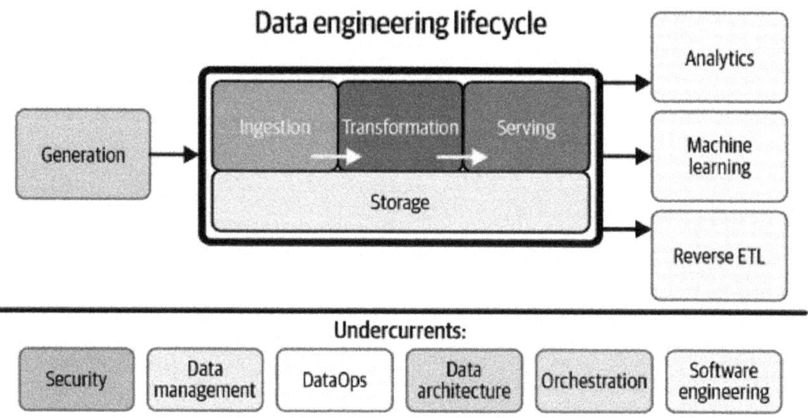

Figure 1.10: Data Engineering Lifecycle

Source: - *(N, 2022)*

The five major components and the ways in which those 6 undercurrents affect each section are shown in the graphic above. The five primary components consist of:

- **Generation**

    Any data source that produces raw data that we can use later is referred to as a source system. They can take many different forms, including online services, RSS feeds, flat files (CSV, XML), transactional databases, and IoT devices.

    The ability to communicate with the upstream stakeholders responsible for managing these sources is essential for Data Engineers (DEs). When there are a lot of systems that a DE needs to know about, this might be challenging.

    Criteria for assessing source systems:

    - What are the essential characteristics of the data source?

    - Length of data persistence, i.e. long-term or quickly deleted?

    - Rate of data generated, i.e. events/size per second/hour?

- Level of consistency, i.e. standards, formatting, cleanliness of data

- Frequency of errors or duplicates

- Timing of data values, i.e. are some later than others?

- What is the schema? How are schema changes communicated (if any)? Is it schema less or fixed schema?

- Frequency of refresh (align to use case)

- Type of tables, e.g. periodic snapshots or update events or logs, etc.

- Performance considerations when reading from data

- Are there data quality checks in place upstream, or is it something you need to address?

- **Ingestion**

  Many businesses frequently experience constraints in source system production and consumption. This is because ingestion can suddenly stop functioning and source systems are maintained outside. A chain reaction affecting your entire lifespan could result from these two components' unreliability.

  Key considerations for evaluating ingestion patterns:

  - What are the use-cases for this data? Can I re-use this data instead of create multiple versions of the same dataset?

  - Destination after ingestion

  - Frequency of access

  - Volume of data during arrival

  - Format of data during arrival

  - Pull vs push

  The choice between batch and streaming-style ingestion is the most contentious one at the moment. By using batch ingestion, data may be taken in pieces that can be defined by size, time period, or criterion. To "stream" data is to "consume" it in a continuous, (almost) real-time manner. The most common data transfer technique has been batch in the past, but streaming is expected to surpass it in the next years. Important factors to consider while deciding between batch and streaming:

- Can downstream storage handle stream flow?

- Micro batch vs Realtime? would minute batches suffice?

- What specific benefits do I gain from using streaming instead of batch?

- What's the difference in cost (money, time, maintenance, downtime, opportunity cost)?

- Would an ML model benefit from continuous training?

Finally, when it comes to ingestion patterns, batch ingestion is the way to go for the majority of use-cases (such weekly reporting or model training), therefore you should only switch to streaming if absolutely necessary because it offers superior trade-offs.

- **Storage**

  Storage plays an important role in every stage of the lifecycle and can manifest in various ways at each stage. There are numerous benefits and drawbacks to consider when selecting storage options. Storage can often take place either on-premises, or in the cloud, the latter of which has seen tremendous growth in popularity in recent years for valid reasons (we won't delve into the advantages of cloud computing here).

  Key considerations for evaluating storage systems:

  - Compatibility with required write and read speeds (+ SLAs)

  - Bottleneck considerations for downstream processes

- How the storage works, i.e. does it prioritise long term storage or frequent and fast reads? Does it support complex query patterns?

- Scalability

- Capturing metadata (schema evolution, data flows, lineage) and managing governance (MDM or golden records + compliance)

- Is it schema less (schema on read) or fixed schema (schema on write)?

Data retrieval patterns will vary greatly across use cases because not all data is the same. Thus, we can classify some data as "hot-data," meaning it is accessed frequently (many times per day or minute), and some data as "cold-data," meaning it is seldom requested and can be archived.

- **Transformation**

The DE is most valuable to the company in this role. If this isn't done, data won't be able to help businesses make decisions. Changes can be as simple or as intricate as the user desires. Data being transformed into various data kinds or standardised formats is a common occurrence on a more fundamental level. After that, we might observe data normalization, schema transformation, or aggregation. Applying business logic to the raw data, however, may result in the most intricate (and significant) change. Another name for this is "data modelling."

Building your data warehouse tables to represent your company unit's structure and operations is one use of data modelling. Establishing what constitutes a customer is one particular business rule that may be necessary to execute. Does this person have more than just an account? Is prior purchase required? Do you have a specific time frame in mind?

Key considerations for transforming:

- Calculating the ROI of the change

- Is the transformation simple and self-isolated?

- What business rules does the transformation support?

Additionally, transformation can occur at other points in the lifecycle; for instance, it can occur during flight when data is consumed and a stream applies specific logic to the final destination. Time stamps and additional fields/calculations are two examples of this.

The final main type of change is feature optimisation. Combining domain knowledge with deep data science understanding to extract and improve features is a common step in getting ready for an ML model.

- **Serving Data**

  The majority of your interactions with other stakeholders will occur during the lifecycle's last stage. Here is where the people farther down the chain genuinely get the rewards of your hard work. We classify the three primary data providing outputs as:

  - **Analytics** — encompasses data analysis conducted on an as-needed basis, released reports or dashboards. Separated into business intelligence, operational analytics, and embedded analytics.

  - **Machine Learning** — include providing data utilised for forecasting or decision-making purposes.

  - **Reverse ETL** — includes re-entering the original data or another system with the changed data as its output.

1.5 Essential Skills and Tools for Data Engineers

The role of the data engineer is extremely varied and entirely dependent on the size of the company and the technology and infrastructure they have. Companies with similar technology stacks can even hire data engineers for two completely different purposes.

That being said, the roles and responsibilities of data engineers typically fall into one of these four core groups:

- Generalists
- Specialists in data storage
- Specialists in programming and pipelines
- Specialists in analytics

Each one of these groups (except for the generalist) corresponds to a specific set of skills and tools that must be mastered to do your job effectively. Knowing which group, you would like to work in can help to focus your learning efforts. Let's go over each of these groups.

Generalists

Data engineer generalists are involved in all aspects of data collection, storage, analysis, and movement. They are typically employed in small companies or companies in the early stages of analytics with small data teams.

The generalist is the hardest role in data engineering, especially for beginners. It can take many years of experience to learn and use the many different tools required by companies.

Specialists in Data Storage

Data engineers specializing in data storage are responsible for setting up and managing databases, data warehouses, and other storage platforms (both in the cloud and on-premise).

Some examples of data storage tools are:

- Relational and non-relational databases like SQL, NoSQL, and PostgreSQL
- Data warehouses like Redshift and Panoply
- Big data systems like Hadoop and Spark
- Cloud-based databases like AWS RDS and Microsoft Azure

These data engineers need a solid understanding of data modeling techniques. The chosen data storage platform should be optimized so that it operates effectively within the budget constraints of the company. Once a database or data warehouse is designed and set up, it needs to be populated. An effective ETL system must also be designed to funnel in the data from possibly many different sources.

Specialists in Programming and Pipelines

Data engineers that focus on programming and pipelines are in charge of developing and overseeing the movement and flow of data. These data engineers must be familiar with many different programming languages and be able to integrate with many different platforms to create data pipelines, automate tasks, and write scripts.

These are the most common programming languages used by data engineers:

- Python
- Java
- C++
- Scala
- Ruby

Specialists in Analytics

Data engineers specializing in analytics work closely with data scientists and other analytics professionals. This means they must understand the tools, techniques, and frameworks used in data-related projects.

Depending on the project, data engineers must be familiar with many areas of data science and analytics, such as:

- Being able to set up and manage ETL tools and pipelines that support these projects (such as Stitch or Airflow)
- Being able to work with big data using tools like Hadoop, Spark, and Kafka

- Knowledge of BI tools and what they require, such as Power BI and Tableau

- Knowledge of ML libraries, such as TensorFlow, Spark, and PyTorch

Data Engineer Requirements

There are usually three main requirements that are considered for data engineer roles:

- Qualifications

- Certifications

- Experience

A bachelor's degree or equivalent is often required of data engineers, while those with experience in mathematics, computer science, engineering, or a similar area sometimes have more extensive education. The role of a data engineer requires a heavy amount of technical knowledge, which is why companies usually require at least a bachelor's degree. While it is also possible to get into data engineering without a technical degree, it is much more difficult, and you will need to do more to prove you have what it takes to do the job.

Certifications are good additions to your resume that can help set you apart from the competition. They prove that you have a good understanding of some of the frameworks or tools required for a job in data engineering.

Qualifications and certifications aside, it is often very difficult to get an entry-level position in data engineering. Companies typically ask for at least a few years of experience in a related field or using the required tools before considering a candidate.

This means you may need to use another data-related role as a bridge to get you into data engineering. It is common for someone to get hired at a company as a software engineer, business intelligence developer, or data analyst and then transfer to a data engineering role after gaining a few years of experience.

Top 5 Data Engineering Skills

Data engineering is an extremely broad and evolving field. There are so many tools, frameworks, and technologies out there that it is almost impossible to know and master all of them. The tools you choose to learn can depend on the company you want to interview for or which data engineer group you fall into.

However, for most data engineering roles, there are five crucial areas you need to develop. If you need somewhere to start, then start with these essential data engineering skills:

1. **SQL Skills**

   The data engineering profession relies heavily on SQL, which is more than just a competence. When you're a SQL expert, you know how to do more than just run queries; you can also traverse and modify complicated datasets utilising different SQL dialects, such NoSQL for unstructured data, PostgreSQL for object-relational database systems, and MySQL for its dependability and versatility. To truly excel in data engineering, developing an in-depth knowledge of these variations is crucial.

   If you're new to Structured Query Language and want to learn the basics, our SQL Fundamentals track is a great place to start. Additionally, we have included a handy SQL Basics cheat sheet that goes over all the necessary queries.

2. **Data Modeling Techniques**

   Data engineers rely on accurate data modelling as a foundation to construct optimised and scalable data warehouses and databases. More than just the layout, it requires knowing how data relates to one another, what limitations exist, and how to scale. An important data engineering talent is mastery of data modelling techniques, since they are necessary for the successful execution of data pipelines, the backbone of data engineering projects.

Our Data Modelling with Power BI course is a great approach to gain a head start in data modelling with tools like Power BI.

3. Python Skills

Python is often regarded as one of the most prominent programming languages. Data cleansing and analysis, as well as integrations and automation, are all under its purview. Additionally, it is a great first language to learn and one of the most adaptable languages overall.

Python is so widely used that it powers the back end of numerous data engineering tools, which in turn make data engineering jobs easy to integrate. If you're interested in studying Python and want to know how to construct efficient data architectures, optimize data processing, and manage large-scale data systems, then you should look into our Data Engineer with Python track.

4. Hadoop for Big Data Skills

Hadoop is one of the most popular specialised systems for working with huge data. It has become a byword for big data due to being a powerful, scalable, and inexpensive technology.

Data engineers are generally tasked with maintaining, testing, analysing, and evaluating enormous data sets that are generated everyday by organisations and people. Our course, Big Data Fundamentals with PySpark, is a great way to begin working with big data.

5. AWS Cloud Services Skills

1. **Core AWS Services & Fundamentals:** AWS offers a range of core cloud services, including compute, storage, networking, and database solutions. Compute services like Amazon EC2 (Elastic Compute Cloud) provide virtual machines with scalable processing power, while AWS Lambda allows for

serverless execution of code without provisioning infrastructure. Containerized applications benefit from Amazon ECS (Elastic Container Service) and Amazon EKS (Elastic Kubernetes Service) for managing workloads efficiently. Storage solutions include Amazon S3 (Simple Storage Service) for object storage, EBS (Elastic Block Store) for persistent block storage attached to EC2 instances, and EFS (Elastic File System) for scalable shared storage. Networking components such as VPC (Virtual Private Cloud) allow users to isolate their cloud resources securely, while Route 53 provides scalable domain name system (DNS) services, and CloudFront enhances content delivery performance. AWS also offers a variety of database solutions, including Amazon RDS (Relational Database Service) for SQL databases, DynamoDB for NoSQL applications, and Redshift for data warehousing.

2. **DevOps & Automation:** Automation is essential in cloud environments for managing infrastructure and streamlining deployments. Infrastructure as Code (IaC) tools like AWS CloudFormation and Terraform enable users to define and provision infrastructure through code. AWS also supports continuous integration and continuous deployment (CI/CD) through services such as AWS Code Pipeline for orchestrating deployments, AWS Code Build for compiling and testing code, and AWS Code Deploy for automating application rollouts. Monitoring and logging services, including Amazon CloudWatch, AWS X-Ray, AWS Config, and AWS CloudTrail, help track system performance, trace requests, enforce compliance, and audit user activities.

3. **Machine Learning & AI on AWS:** AWS provides a robust ecosystem for artificial intelligence and machine learning. Amazon Sage Maker allows users to build, train, and deploy machine learning models at scale with automated

hyperparameter tuning and Jupyter Notebook integration. AWS also offers AI-powered services like Amazon Rekognition for image and facial analysis, Amazon Polly for text-to-speech conversion, Amazon Lex for chatbot development, and Amazon Comprehend for natural language processing. These services empower businesses to integrate machine learning into their applications with minimal effort.

4. **Data Analytics & Big Data:** Data analytics and big data processing are critical for deriving insights from vast amounts of information. AWS enables batch and real-time analytics through services such as AWS Glue for ETL (Extract, Transform, Load) processes, Amazon Athena for serverless SQL querying of S3 data, and AWS Lake Formation for managing data lakes. Amazon Kinesis and AWS Managed Kafka provide real-time data streaming and processing capabilities, while Amazon QuickSight delivers business intelligence and visualization tools to help users make data-driven decisions.

5. **Security & Compliance:** Security is a top priority in cloud computing, and AWS offers extensive security controls. Identity and Access Management (IAM) allows for granular permissions and role-based access control, while AWS Organizations simplifies multi-account management. AWS provides advanced security services, including AWS WAF & Shield for web application firewall and DDoS protection, AWS GuardDuty for threat detection, and AWS Security Hub for centralized security monitoring. Compliance management tools like AWS Artifact and the AWS Well-Architected Framework help businesses adhere to industry regulations and best practices for building secure cloud environments.

6. **Cloud Architecture & Best Practices:** Building resilient and cost-effective cloud solutions requires following best practices in cloud architecture. AWS promotes high availability and fault

tolerance through Multi-AZ (Availability Zone) deployments, Auto Scaling Groups, and Elastic Load Balancers (ELB) to distribute traffic across multiple instances. Cost optimization strategies include leveraging AWS Budgets, AWS Cost Explorer, and using Spot Instances or Savings Plans for reducing compute expenses. AWS also supports multi-region deployments and hybrid cloud strategies through AWS Outposts and AWS Direct Connect, ensuring seamless integration with on-premises environments.

7. **AWS Certifications & Learning Path:** To validate expertise in AWS cloud services, professionals can pursue certifications that align with their career goals. Associate-level certifications such as AWS Certified Solutions Architect – Associate, AWS Certified Developer – Associate, and AWS Certified SysOps Administrator – Associate cover fundamental skills. Advanced certifications like AWS Certified Solutions Architect – Professional and AWS Certified DevOps Engineer – Professional provide in-depth knowledge of designing scalable solutions and implementing automation strategies. AWS also offers specialized certifications in Machine Learning, Security, and Big Data to focus on specific domains.

8. **Hands-on Projects & Real-World Applications:** Practical experience is key to mastering AWS services. Hands-on projects such as deploying a serverless API using Lambda and API Gateway, setting up an Auto Scaling web application with EC2 and ALB, building a data pipeline using AWS Glue and Athena, and training machine learning models in SageMaker can reinforce knowledge and provide real-world experience. Implementing IAM security best practices, managing a Kubernetes cluster on EKS, and optimizing cost and performance using AWS monitoring tools further enhances cloud expertise.

1.6 Data Engineering vs. Data Science

Both Data Scientists and Data Engineers play important roles in the collection, analysis, and utilization of data, but their responsibilities, skill sets, and objectives are distinct. Understanding the differences between a Data Scientist and a Data Engineer is essential for organizations seeking to build robust data teams and for individuals considering careers in these fields.

Data Scientist: The main goal of a data scientist is to assist organisations in making well-informed decisions by analysing and understanding complex data. They frequently collaborate directly with company stakeholders to comprehend certain objectives and enquiries, examine data patterns, and develop models to forecast future results.

Roles & Responsibilities of Data Scientists

- **Data Analysis and Interpretation:** Data scientists analyse data using computers and statistical methods. They interpret data trends and patterns to provide actionable insights.

- **Model Building:** They create algorithms for ML and predictive models to foretell how people would act in the future.

- **Data Visualization:** Assisting stakeholders in understanding data by developing visual representations of findings.

- **Experimentation:** Creating and carrying out experiments to verify models and test theories.

- **Reporting:** Summarizing findings in reports and presentations to inform business strategies.

Data Scientist Career Path:

- **Junior Data Scientist:** Typically focuses on basic analysis and assisting with model development.

- **Data Scientist:** Takes on independent projects, analyzes data, and builds models for business needs.

- **Senior Data Scientist:** Leads modeling initiatives and may specialize in fields like deep learning or NLP.

- **Machine Learning Engineer/Research Scientist:** Focuses on advanced model development and deployment.

Data Engineer: A Data Engineer, on the other hand, is responsible for the design, construction, and maintenance of the data infrastructure. They create robust systems to gather, store, and process data, ensuring data pipelines are efficient, reliable, and scalable.

Roles & Responsibilities of Data Engineer

- **Data Architecture Design:** Making ensuring data flows and is stored efficiently through designing data systems and pipelines.

- **Data Pipeline Development:** Construction and maintenance of data pipelines for the transfer of data between data storage and processing systems and other sources.

- **Database Management:** Database administration and optimisation for optimal data accessibility, performance, and integrity.

- **ETL Processes:** Constructing ETL procedures to get data ready for analysis.

- **System Integration:** Making sure data flows freely across systems and integrating several data sources.

Data Engineer Career Path:

- **Junior Data Engineer:** Focuses on data cleaning, basic pipeline construction, and database management.

- **Data Engineer:** Designs and maintains data architectures and ETL processes.

- **Senior Data Engineer:** Manages complex data infrastructures and optimizes systems for scale.

- **Data Architect:** Designs the overall data ecosystem, choosing platforms and tools for optimal performance.

- **Chief Data Architect:** Responsible for an organization's data strategy and architecture decisions.

Difference between Data Scientist and Data Engineer

Aspect	Data Scientist	Data Engineer
Primary Focus	Analyzing and interpreting complex data to provide insights	Designing, building, and maintaining data infrastructure
Goals and Objectives	Predictive Analytics, Decision Support, Optimization, Innovation	Data Accessibility, Data Quality, System, Efficiency, Scalability
Required Skills	<ul><li>Programming (Python, R, SQL)</li><li>Statistical Analysis</li><li>Machine Learning</li><li>Data Visualization</li><li>Big Data Tools (Hadoop, Spark)</li></ul>	<ul><li>Programming (Python, Java, Scala, SQL)</li><li>Data Warehousing</li><li>ETL Tools</li><li>Big Data Tools (Hadoop, Spark, Kafka, Flink)</li></ul>
Tools and Technologies	<ul><li>Python, R, SQL</li><li>TensorFlow, scikit-learn, Keras</li><li>Tableau, Power BI, matplotlib</li><li>Hadoop, Spark</li></ul>	<ul><li>Python, Java, Scala, SQL</li><li>Amazon Redshift, Google Big Query, Snowflake</li><li>Apache NiFi, Talend, Informatica</li><li>MySQL, PostgreSQL, MongoDB, Cassandra</li></ul>

Aspect	Data Scientist	Data Engineer
Educational Background	Statistics, Mathematics, Computer Science	Computer Science, Software Engineering, Data Management
Collaboration	Works with Data Engineers to define data needs and quality, uses data infrastructure built by Data Engineers	Works with Data Scientists to provide reliable data pipelines, Builds and maintains the infrastructure used by Data Scientists
Nature of Work	Analytical	Engineering and Technical
Problem-Solving Approach	Hypothesis testing and experimentation	Systematic and architectural design
Typical Employers	Research organizations, financial institutions, Technology firms	Tech companies, large enterprises with data needs, Data-focused start-ups

1.7 Challenges and Opportunities in Data Engineering

There are many opportunities and problems in the ever-changing field of data engineering. Understanding these can help professionals navigate the landscape effectively.

Challenges in Data Engineering

- **Data Discovery**

  Data discovery is figuring out what kinds of data are required, what systems are already in place, and how to integrate all this data into effective pipelines. In order to solve business challenges, model your data, and discover the value in your data, this first step is essential.

The difficulty of data discovery, however, grows as organisations expand and interact with multiple suppliers and various data sources. The integration process is like putting together a jigsaw puzzle because each source has its own unique formats, schemas, and standards. Integrating data securely and efficiently necessitates not just a solid technical grasp but also a keen awareness of how to adhere to different regulations.

Data specialists must be meticulous and explicit throughout this stage to prevent issues in the future. Teams risk coming face to face with production-level problems when they rush to begin data initiatives without properly planning this area.

- **Data Silos**

   A major obstacle in the world of contemporary data management is data silos. An all-too-common problem in modern businesses is data silos, which occur when different parts of an organization's systems store data in isolated locations. A disjointed data ecosystem with little data integration is the result of no vendor offering a one-size-fits-all solution that can manage all parts of a company. The effects of data silos are far-reaching and detrimental to businesses. For instance, if the marketing and sales teams are using different databases, they may miss out on sales prospects because the full customer journey cannot be analyzed.

   In addition, operational expenses are escalated due to data maintenance activities being duplicated across several teams. Data quality and confidence are eroded by inconsistency and inaccessibility. There needs to be an immediate effort to promote data integration and accessibility at all levels of an organisation because this mistrust of data can severely limit strategic decision-making and growth.

- **Data Quality**

  Data quality is paramount. The data value chain as a whole is at risk in the absence of comprehensive, accurate, relevant, and consistent high-quality data. Data validation and cleansing procedures must be rigorously implemented by data engineers because the quality of data has a direct impact on the insights obtained from it.

  A lack of adherence to stringent data standards can result from a number of sources, including manual data input methods or third-party vendors. Data accuracy, completeness, and consistency are at stake, and these problems need finding and fixing faulty, duplicate, or incomplete data.

  The key to improving data quality is not to discover a silver bullet, but to make it an integral part of your data governance strategy from the get-go. To ensure data quality criteria are met, which promotes data-driven decision-making, it is vital to use advanced tools and techniques. The objective is to salvage data lakes from the brink of becoming useless data swamps and turn them into useful resources.

- **Data Compliance and Monitoring**

  Compliance with different data regulations has grown increasingly complex and important as on-premises to cloud systems have been transitioned to. To safeguard sensitive financial and personal information, stringent compliance with regulations such as GDPR and HIPAA is required. The difficulty lies in the fact that it is not enough to simply comply with these rules; constant monitoring and auditing of data is required.

  Particularly difficult tasks today include data auditing and monitoring. There is a clear lack of maturity in the methods used for auditing and monitoring data as it migrates to the cloud and the amount and variety of data continue to increase. The data engineering landscape is lacking in this area, which impacts data quality, compliance, and security.

Without a defined procedure for data audits and monitoring, businesses run the danger of non-compliance fines, security lapses, and problems with data quality. To overcome this obstacle, a concentrated effort must be made to create increasingly complex frameworks and procedures for data governance, auditing, and monitoring. Only then will companies be able to guarantee the security, compliance, and quality of their data assets.

- **Data Masking and Anonymization**

  Comprehensive data masking and anonymisation procedures are necessary for data protection. These methods are essential for preventing unwanted access to private data while maintaining data usability for processing and analysis. The objective is to guarantee that data confidentiality and privacy are preserved without sacrificing the usefulness of the data, whether this is accomplished by data shuffling, data scrambling, or the use of synthetic data.

- **Business Value Generation**

  Producing measurable business value is the end goal of every data endeavour. This necessitates upfront transparency regarding the possible physical and intangible commercial benefits of data projects. The primary focus of executives and decision-makers is the financial effect and growth drivers of data projects.

  Making sure data projects are connected to delivering value for the company and not merely technical exercises is the real difficulty. Funding and backing for data projects depend on this congruence.

- **Maintaining Data Pipelines**

  A data pipeline is a vital component of any data infrastructure as it ensures the efficient and connected transfer of data from its point of origin to its final destination. In order to guarantee the health and dependability of these pipes, data engineers must pay close attention to their construction and maintenance, which is a fundamental obligation. As part of this process, data transfers are continuously monitored, bottlenecks are identified and resolved swiftly, and optimisations are implemented strategically to prevent disruptions.

Improving reliability and efficiency of these systems requires following best practices including adding redundancy, making sure there is fault tolerance, and creating strong recovery plans. In addition, regular maintenance aims to decrease downtime, improve scalability, and refine data flow, making sure that pipelines efficiently move and process data. Companies can guarantee a steady stream of high-quality data through these pipelines, which in turn helps data analysts and scientists draw conclusions and take actions based on that data.

- **Minimizing Human Error**

  Data quality is greatly endangered by human error, which is an inherent part of data administration. Even small mistakes can compromise the trustworthiness and accuracy of data systems. It is critical to adhere to thorough governance principles and establish strict data validation procedures in order to reduce this risk. Data security, quality, and mistake-free status necessitate such actions, as does the reduction of human error.

  Data that is correct, complete, and consistent may be maintained by organisations by strictly following the right validation methodologies and governance rules. This allows for the extraction of important insights. This, in turn, facilitates data-driven strategy development and informed decision-making, both of which are critical to an organization's success. The possible detrimental effects of human error on data analysis and decision-making procedures can be greatly reduced by a dedicated adherence to these procedures, improving the data's overall security and quality.

- ## Choosing the Right Tools and Technologies

  Data engineers find it challenging to stay up to date and select the appropriate tools for their organization's requirements due to the rapid pace of change in data engineering technologies. It can be challenging to find tools that complement the company's current operations, are reasonably priced, and offer appropriate support, particularly for those accustomed to less sophisticated, less efficient approaches like spreadsheets or basic software analytics. Over-reliance on some systems' basic reporting features can slow down work and lead to data silos, which makes it difficult to fully comprehend the data.

  The backdrop of assessing sales success across different product lines provides a real-world example. There is often a lack of consistency in the formats and structures used by various systems when it comes to sales data. This is especially true in many companies' CRM platforms, sales automation tools, and finance software. Companies frequently resort to manual data consolidation using spreadsheets due to the difficulties of integrating these varied data sources when trying to analyse overall sales performance, segment profitability, or the effectiveness of individual salespeople. The procedure is inefficient and may be inaccurate as a result of the high risk of mistakes and the large amount of time and resources consumed. An integrated data stack, bolstered by robust business intelligence (BI) tools, is essential in light of these problems. To gain real-time insights into trends, performance indicators, and improvement opportunities, businesses need the correct BI solution to automate data aggregation and analysis across many sources.

  The data engineering landscape will keep changing as we go into 2024 and beyond, bringing with it fresh problems and opportunities for innovative approaches to solving them. To turn these obstacles into opportunities for development and innovation, data engineers must be able to keep updated, embrace innovation, and foster a culture of continuous improvement.

Opportunities in Data Engineering

- **Data Engineer**

  Designing, constructing, and managing data pipelines and infrastructure is the responsibility of a data engineer, who is considered the foundational function in data engineering. You will need to deal with a variety of data sources, perform ETL operations to extract, process, and load data, and check that the data is reliable and of high quality.

  For organisations to make data-driven decisions, data engineers are crucial in planning, building, and managing the data infrastructure.

  Your main goal should be to build data pipelines that efficiently integrate data from various sources and convert it into a format that can be stored and analysed.

  You will be responsible for ensuring the availability, security, and performance of data across various cloud platforms and databases. You will be an integral part of the team that works directly with analysts, data scientists, and other stakeholders to ensure that they have access to accurate data for modelling and insights.

  Improving your data engineering abilities will not only make you look nice, but will also help you grasp the concepts much better.

  Data engineering is a thriving field that can lead to exciting new chances for personal and professional advancement thanks to the ever-increasing volume of data and the growing significance of data-driven tactics.

- **Big Data Engineer**

  Big data engineers are experts in managing massive datasets through the use of distributed systems such as Spark and Apache Hadoop.

For effective processing and analysis of large data sets, you should develop and deploy big data solutions.

Your primary focus will be on creating and overseeing data systems capable of managing large amounts of both structured and unstructured data. Modern tools that make use of distributed computing and parallel processing will be at your disposal as you handle data efficiently.

Your knowledge in data pipeline optimisation, data integration, and data quality assurance can help organisations gain important insights and make decisions based on data.

A career as a big data engineer presents endless opportunity to tackle challenging challenges and contribute to revolutionary advancements, as big data continues to alter industries and drive innovation.

- **Cloud Data Engineer**

  Data solutions on the cloud are the speciality of cloud data engineers. You will mainly be responsible for developing and overseeing storage systems and data pipelines in AWS, Azure, or GCP cloud environments.

  Effective data extraction, transformation, and loading will be your speciality as you collaborate with a wide range of cloud-based data services and technologies. You will facilitate data scientists' and analysts' access to and analysis of data by working with cross-functional teams to guarantee data availability, security, and dependability.

  When it comes to the ever-changing digital landscape, your knowledge of cloud data engineering is going to be crucial in assisting organisations in making smart decisions, fostering innovation, and maximising the capacity of their data.

- **Data Architect**

  An organization's data architecture is the blueprint for its data systems and infrastructure as a whole. Building a unified and effective data ecosystem requires careful planning of data storage, data movement, and integration across many data platforms.

  Understanding business objectives, developing data models, and integrating data from many sources will be your main responsibilities as you strive to build a unified and extensible data ecosystem.

  You will be responsible for coordinating with data analysts, stakeholders, and engineers to guarantee the safety and accuracy of all data while also making sure it complies with all applicable laws and industry standards.

  Your knowledge and experience can help businesses maximise their data assets, develop data-driven strategies, and make smart decisions that lead to long-term success by improving data flow and integrating data seamlessly.

- **Data Warehouse Engineer**

  Data warehouse engineers are experts in creating and managing databases designed for reporting and analysis. Tools like SQL databases and data warehousing platforms are part of your toolbox.

  Your primary responsibilities will centre on building data pipelines, optimising data models, and maintaining data quality and consistency in the data warehouse.

  You will work closely with data architects, ETL developers, and data analysts to help turn raw data into meaningful insights. This will ensure that stakeholders have the knowledge they need to make data-driven decisions.

  An organization's ability to gain growth and competitive advantage depends on your ability to lay the groundwork for their data warehousing, SQL database, and data integration efforts.

- **Data Governance and Security Specialist**

  Data privacy, security, and compliance are the key concerns of data governance and security experts. Protecting the confidentiality, availability, and integrity of a company's data is an important responsibility that you will have. Your major duty will be to regulate the use, access, and administration of data by creating and enforcing rules, processes, and best practices for data governance.

  Protect sensitive information and ensure compliance with data requirements by implementing data security measures, encryption methods, and access controls in collaboration with cross-functional teams. In order to reduce risks and ensure data integrity, you will be responsible for doing data audits, keeping an eye on data usage, and handling security incidents.

  Your knowledge and experience will be crucial in establishing a data security and trust culture, which is essential in this age of increasing cyber-attacks and data breaches. This will allow organisations to responsibly handle data, safeguard customer data, and keep their credibility and reputation in this data-driven world.

- **Data Consultant**

  Experts in data analysis can be found working for themselves or for larger consulting companies. Its primary function is to assist businesses in enhancing their data infrastructure and operations by providing advice and assistance on data engineering best practices.

  Your primary objective should be to evaluate their current data architecture, procedures, and analytics requirements in order to pinpoint potential optimisation and enhancement opportunities. Collaborating with stakeholders, you will use your expertise in data science, data engineering, and business intelligence to create data-driven solutions that meet their goals.

In addition to providing guidance on data security, compliance, and governance, you might provide a hand with data modelling, data visualisation, and data integration initiatives.

Leading data-driven change, assisting organisations in making informed decisions, and accomplishing business objectives through data utilisation will require your adaptability and the capacity to explain intricate technical ideas to stakeholders without a technical background.

As a Data Consultant, you'll get to put your skills to use across a wide range of sectors and projects, allowing you to learn new things and make a real difference in solving complex business problems.

- **Data Integration Specialist**

  Integrating data from a wide variety of internal and external sources is the speciality of data integration professionals. Making sure data is correctly linked and easily accessible is crucial.

  Making sure data can flow freely and easily between systems and applications is going to be your main focus as you plan, create, and execute data integration solutions.

  Ensuring data consistency and quality is your top priority as you deal with a variety of data formats, APIs, and integration technologies to do ETL. You will work on solving difficult data integration problems and enabling effective data sharing and analysis in a team including data engineers, database administrators, and company stakeholders.

  Simplifying data procedures, improving data workflows, and enabling businesses to fully utilise their data for data-driven decision-making and commercial success will all be made possible by your proficiency in data integration.

Multiple Choice Questions (MCQs)

1. **Which of the following is a significant milestone in the evolution of data engineering?**

    a. Development of relational databases

    b. Introduction of data lakes

    c. Emergence of cloud computing

    d. All of the above

2. **What is the primary responsibility of a data engineer in a modern enterprise?**

    a. Building and maintaining data pipelines

    b. Analysing data to generate insights

    c. Designing machine learning models

    d. Creating marketing strategies

3. **What does ETL stand for in data engineering?**

    a. Extract, Transform, Load

    b. Evaluate, Transfer, Learn

    c. Execute, Transfer, Load

    d. Extract, Translate, Learn

4. **Which phase in the data engineering lifecycle involves the process of moving raw data from different sources to a data warehouse?**

    a. Data Collection

    b. Data Integration

    c. Data Analysis

    d. Data Storage

5. **Which of the following tools is commonly used for data storage in data engineering?**

   a. Jupyter Notebook

   b. Apache Hadoop

   c. TensorFlow

   d. Power BI

6. **Which of the following is the key difference between data engineering and data science?**

   a. Data engineers build data pipelines, while data scientists focus on interpreting the data.

   b. Data engineers work with machine learning models, while data scientists design them.

   c. Data engineers primarily focus on visualization tools, while data scientists focus on database management.

   d. There is no difference between data engineering and data science.

7. **Which of the following is a major challenge faced by data engineers?**

   a. Managing the quality of data

   b. Interpreting data insights

   c. Building predictive models

   d. Designing marketing campaigns

8. **Which of the following programming languages is essential for data engineering?**

   a. Python

   b. JavaScript

   c. Ruby

   d. HTML

9. **In the data engineering lifecycle, which phase involves cleaning and transforming data for analysis?**

    a. Data Collection

    b. Data Transformation

    c. Data Storage

    d. Data Visualization

10. **What is one of the key opportunities in data engineering?**

    a. Automating data cleaning and transformation processes

    b. Developing complex machine learning models

    c. Performing in-depth data analysis

    d. Designing marketing campaigns

Answer

1	2	3	4	5	6	7	8	9	10
D	A	A	B	B	A	A	A	B	A

Chapter 02

DATA ACQUISITION AND STORAGE

2.1 Understanding Data Sources

The components of a Data Acquisition System, sometimes shortened to DAQ, include sensors, measuring devices, and a computer. An understanding of electrical or physical phenomena relies on its ability to collect and process crucial data. Tasks such as evaluating the effectiveness of obtaining desired levels by monitoring the heating coil temperature are greatly aided by this technology. Data acquisition, often called data collection, is dependent on dedicated software that swiftly records, analyses, and stores data. It allows for thorough analysis for scientific or engineering objectives, which is useful for scientists and engineers. Different measurement needs can be met by utilising one of the several data gathering systems that are available in both portable and remote versions. Remote systems are great for long-distance measurements and offer more flexibility in data collecting, whilst handheld devices are better suited for direct subject involvement.

What Does a Data Acquisition System Measure?

Many different parameters, mostly obtained from analogue signals, can be measured with the use of data acquisition systems. The digitisation of these measurements allows for their processing by computers, and they play an important role in a number of applications.

This data collecting device measures a wide variety of variables, including but not limited to: current, voltage, strain, frequency, pressure, temperature, distance, vibration, angles, digital signals, weight, and beyond. Expert sensors or modules can be utilised to measure particular parameters with precision and efficiency.

In most cases, voltage is used as the starting point for measuring various other parameters, such as temperature or displacement. Data acquisition modules, in conjunction with suitable sensors or transducers, allow for the effective measurement of practically any desired parameter. Because of this flexibility, data gathering systems can be tailored to meet the needs of various measurement tasks and can be further specialised as needed.

Importance of Data Acquisition Systems

There are multiple reasons why data gathering systems are so important in so many different sectors:

Accurate Data Collection: There is less room for human mistake and more assurance of data integrity because to the facilitation of precise and consistent data collection from a variety of sensors and sources.

Real-Time Monitoring: Data acquisition systems reveal process insights in real time. Better safety and operational efficiency are the results of being able to react quickly to changing conditions.

In industrial and manufacturing contexts, data collecting systems are crucial for quality control. The quality of the items is ensured by closely watching the parameters.

Research and Development: They successfully support research endeavours by providing essential data for experiments, simulations, and the development of new technologies and products.

Environmental Monitoring: Environmental studies rely heavily on data collecting. It is useful for assessing environmental hazards, climatic conditions, and the effects of human actions on ecosystems.

These systems are crucial when it comes to medical applications. Their careful attention to a patient's vital signs helps doctors make correct diagnoses and advances medical technology and treatment. The ability for machines and processes to function well without human intervention is dependent on the data collected in automated systems. This essential part of automation guarantees that everything runs smoothly and performs at its best.

An organization's ability to save and retrieve data is critical for maintaining access to historical data for analysis, compliance, and auditing. This process helps organisations analyse their past more effectively by allowing for fast access to information and stress-free storage.

Energy Management: Data collection systems are vital in the energy sector for tracking generation, transmission, and consumption of power. As a result, conservation efforts are encouraged and energy management is made more efficient.

Components of Data Acquisition System

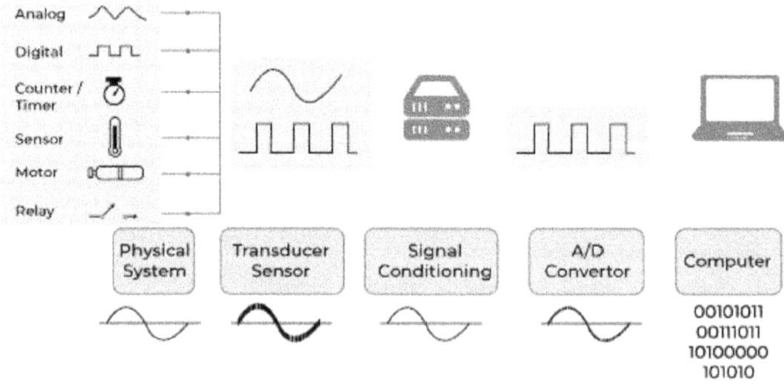

Figure 2.1: Components of Data Acquisition System

Source: - *(Geeksforgeeks, 2024a)*

- **Sensors:** instruments that collect data regarding physical or environmental factors like temperature, pressure, or light intensity.

- **Signal Conditioning:** The raw data from the sensors is pre-processed to remove noise and scale it correctly in order to make sure the measurements are accurate.

- **Data Logger:** Continually recording and storing the conditioned data, either through hardware or software.

- **Analog-to-Digital Converter (ADC):** Processes analogue sensor signals by transforming them into digital data.

- **Interface:** Provides a means of controlling and transferring data from the data collecting system to a separate computer or controller.

- **Power Supply:** Supplies the energy required to run the system and its sensors.

- **Control Unit:** Tasks like triggering, timing, and synchronisation are part of the data acquisition system's overall operation that must be monitored by the manager.

- **Software:** Gives users the ability to set up the system, keep tabs on it, and analyse the data it gathers.

- **Communication Protocols:** Data communication refers to the process of sending and receiving information between a system and other devices or networks.

- **Storage:** Memory cards, hard discs, or cloud storage are some of the available methods for storing recorded data. These offer options for both short-term and long-term storage.

- **User Interface:** Users are able to efficiently engage with and manage the data collecting system through this system.

- **Calibration and Calibration Standards:** Regular calibration against established standards checks the system and sensors for accuracy.

- **Real-time Clock (RTC):** To guarantee synchronised timestamping and data gathering, accurate timing is maintained.

- **Triggering Mechanism:** Events or situations that have been predetermined trigger the data collecting process.

- **Data Compression:** Data compression is an ongoing process that aims to make acquired data more manageable for distant or resource-constrained applications.

Applications of Data Acquisition System

- Collecting sensor data in real-time to enhance manufacturing process efficiency and quality control.

- Ensuring compliance with regulations and safeguarding health by monitoring levels of water pollution and air pollution.

- Gathering information from many scientific domains to back up analyses and test hypotheses.

- Timely medical actions in hospitals depend on continuously monitoring patient data and vital signs.

- Conducting emissions, safety, and performance evaluations of vehicles throughout their development.

- Data collection is vital for ensuring safety and achieving optimal performance during testing of aircraft and spacecraft.

- The optimisation of resource usage is achieved by the monitoring of energy consumption in buildings and industry.

- Conducting safety and maintenance evaluations of infrastructure, such as buildings and bridges.

- Logistics and safety can be improved by monitoring vehicle locations, speeds, and conditions.

- It is vital for grid management to guarantee the consistency and dependability of power generation, distribution, and consumption.

Basic Types of Data Acquisition Systems

The following are examples of the most fundamental data acquisition systems:

1. **Digital Data Acquisition Systems**

   Sensors, instruments, and other sources of data must be collected and processed by means of digital data acquisition systems (DAS). They provide advantages in a variety of fields. These technologies make sure everything is accurate by digitalising analogue signals. Reducing data loss when storing and transmitting is important. Analogue DAS often use components including ADCs, microcontrollers, and data storage units to offer control and analysis in real time. This greatly improves the dependability and efficiency of operations.

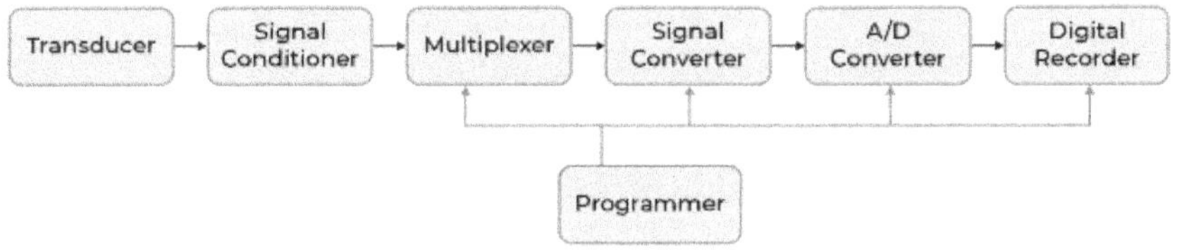

Figure 2.2: Digital Data Acquisition Systems

Source: - *(Geeksforgeeks, 2024a)*

In addition to being able to integrate with computer-based management and monitoring systems with ease, digital DAS provide adaptability in handling different kinds of sensors. As a result, they have found use as instruments in several domains, including but not limited to: research, industrial automation, medical monitoring, and environmental investigations. Their ability to effectively collect, analyse, and disseminate data is useful for improving processes and making well-informed choices in many domains.

2. Analog Data Acquisition Systems

The ability to transform analogue signals from the actual world into digital data for processing and analysis is where Analogue DAS come in handy. The sensors in these systems measure voltage or current in analogue form, and the signal conditioning circuitry in them filters, amplifies, and preprocesses the signals. These analogue signals are transformed into a format that computers or microcontrollers can store and analyse with the help of ADCs.

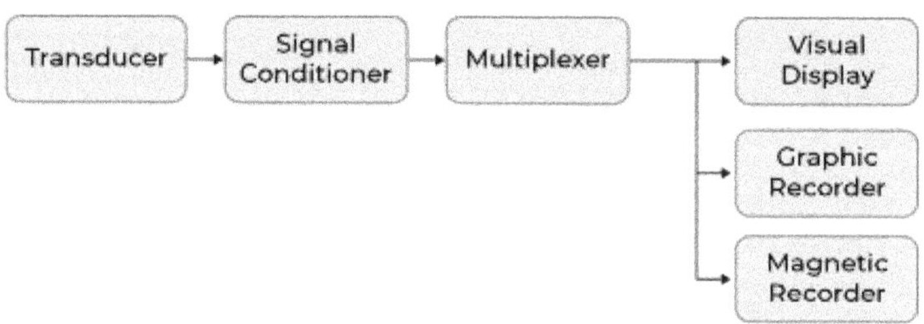

Figure 2.3: Analog Data Acquisition Systems

Source: - *(Geeksforgeeks, 2024a)*

Analogue DAS is useful in many different areas, including medical diagnostics, scientific research, environmental monitoring, and industrial automation. Organisations can regulate operations, monitor vital metrics, and make informed choices using data that is accurate, precise, and reliable. In order to facilitate progress in a variety of sectors, these systems serve as adaptable instruments that connect the real world with digital data analysis.

Data Acquisition Cards and Modules

Rack modules that are fitted with numerous cards allow data collection systems to support varied measurement operations. These cards must be in perfect harmony with the mechanical and electrical connections of the system. Modules are readily available from different vendors and standardised rack systems are often used, which makes the selection process easier for consumers. Data capture configurations are made more adaptable and convenient by this versatility.

Data Acquisition Software

An essential component of effective data management is data collecting software. The software may be tailor-made in several programming languages to meet individual requirements, or you can choose it from a variety of pre-made packages. There are less maintenance concerns with proprietary data collecting software since it uses pre-developed and validated solutions.

Although there are expenses associated with software maintenance, they are usually less than the costs of operating in-house solutions. As a result, a lot of businesses choose to buy data collecting software and modify it to meet their unique testing needs.

Data Acquisition Transducer Signals

The acquisition of data is greatly aided by transducers. The process of transforming energy into electrical impulses is carried out by them. The parameters being measured determine the variation in the output signal generated by these devices. The output type of data collecting systems may be either digital or analogue, and this is a typical way to identify them.

Considerations When setting up a Data Acquisition System

It is crucial to keep in mind the following elements while setting up a data gathering system:

- **Sensor Selection:** Careful selection of the right sensors or transducers is required to correctly gather the required data. In order to make a well-informed choice, factors like measurement range, resolution, and sensitivity must be taken into account.

- **Signal Conditioning:** It is possible to improve the acquired data quality by using signal conditioning methods. Possible adjustments may include amplifying, filtering, or other related processes.

- **Noise Reduction:** To reduce the likelihood of data distortion caused by noise interference, some measures may be put into place. Using strategies like as shielding, grounding, and filtering may be a successful strategy.

- **Calibration:** It is crucial to calibrate measurement devices and sensors on a regular basis to ensure their accuracy and dependability.

- **Data Storage:** The right way to store data is the subject of the next decision. Options including on-premises storage, cloud-based solutions, or a mix of the two should be thought about.

- **Data Transmission:** It is crucial to devise a strategy for data transmission from far sensors to a main system if that is relevant. Methods of communication that are both reliable and secure should be given precedence in this approach.

- **Power Supply:** Making ensuring the sensors and data collecting devices have a consistent and dependable power source is crucial to avoid data loss or system malfunctions. This will ensure that functioning is not disrupted.

- **Environmental Conditions:** The data collecting system's operating environment must be carefully considered. Protecting the equipment from elements such as high humidity, extreme temperatures, and other environmental hazards is crucial.

- **Data Processing:** Specify the methods for processing, analysing, and visualising data. Choose the right data analysis software tools and techniques.

- **Data Security:** Protect sensitive data from breaches and unwanted access by putting security measures in place. Perhaps access limits and encryption are required.

- **Scalability:** Future data growth or sensor additions should be taken into consideration while designing the system to ensure scalability.

- **Regulatory Compliance:** If the data collection system will be handling sensitive or regulated information, it is very important that it adheres to all applicable industry standards and laws.

Data Acquisition Signal Used

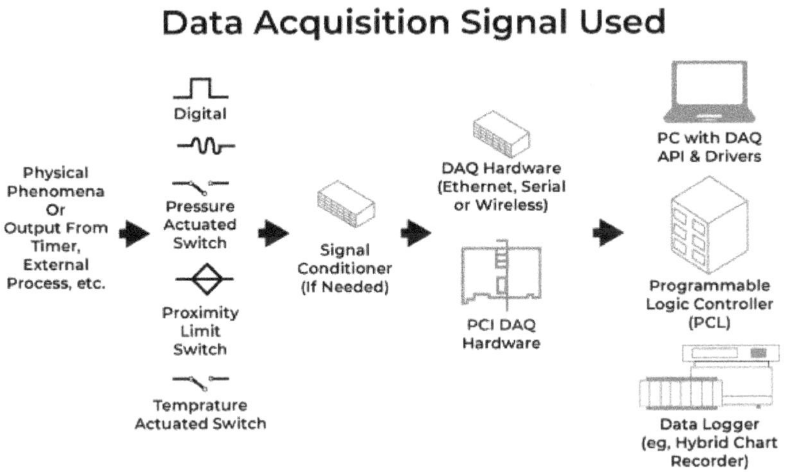

Figure 2.4: Data Acquisition Signal Used

Source: - *(Geeksforgeeks, 2024a)*

Certainly, let's provide more details on each of the signal types used in data acquisition:

1. Voltage Signals

The potential difference between two places in a circuit may be measured using voltage signals. Analogue quantities, such as voltage, taken from sensors, transducers, or electronic devices may be monitored with the use of these electrical measurements. Voltage signals are crucial in many applications where precise electrical measurements are required, including electronics testing, power monitoring, and environmental sensing.

2. Current Signals

Essential for measuring and monitoring electrical currents are the current signals seen in a circuit. Important applications that they are involved in include controlling electric motors, managing batteries, and guaranteeing electrical safety.

3. Power Signals

Power signals are useful resources for controlling energy use effectively. Their data on voltage, current, and other variables allows us to track and improve power consumption in many environments, including buildings, industrial operations, and electrical networks.

4. Thermocouples

There is a broad variety of industries that use thermocouples, which are temperature sensors that generate voltage in response to changes in temperature between two distinct materials. Among them, you may find climate monitoring initiatives, scientific research projects, and industrial operations.

5. Resistance

The evaluation of components or materials is a part of resistance measurements. The method confirms the reliability of electrical connections, which is a crucial aspect of electronics. Understanding material qualities like conductivity and resistivity may be aided by resistance measurements in the field of materials science.

6. Strain Gauge Bridges

Objects subjected to mechanical stress may have their deformation or strain measured using strain gauges. To precisely measure changes in resistance produced by strain, strain gauge bridges are employed in engineering applications. Stress testing, structural analysis, and load monitoring are all made easier with this useful tool.

7. **Digital Signals**

Digital signals, which are usually binary, represent discrete states. Both high and low may be used to describe these situations. They are crucial for keeping an eye on and regulating digital systems and devices, such as microcontrollers, switches, and digital sensors. Moreover, digital signals are ubiquitous in automation, telecommunications, and computer systems.

Advantages of Data Acquisition Systems

- Data capture systems exhibit exceptional accuracy in scenarios when it is fundamental. In situations when high degrees of precision are required, these systems may provide readings that are very precise.

- Data may be collected and monitored in real-time with the help of real-time monitoring systems. This allows for the rapid detection of irregularities and makes fast decisions possible.

- Versatility: Data collection systems are quite versatile and may be used for a variety of activities. The versatility and interoperability of these systems with a wide variety of sensors are shown in scientific research endeavours as well as industrial process control activities.

- Data storage often includes methods for collecting and archiving information, which allows users to access past data and see trends over time.

Disadvantages of Data Acquisition Systems

- The procurement of specialised gear and software makes the process of setting up a data collecting system potentially costly. These essential components might cause the total cost to be high.

- Setup and maintenance of these systems may be challenging, requiring knowledge of both software and hardware components. Intricate in nature, the setup and maintenance operations are complicated.

- Ensuring the proper integration of various sensors and equipment may be challenging due to compatibility difficulties.

- The storage of sensitive information in these systems raises concerns about data security due to the lack of appropriate cybersecurity measures to reduce risks.

- The overall cost and effort required to establish a system might be increased due to maintenance needs, which are important for effective performance.

Methods of Data Acquisition Systems

Presented below are a few approaches to data collection systems:

1. **Bit-Stream Disk-to-Image File**

    - **Purpose:** Used to create backups of original discs for the purpose of preserving data during forensic investigations.

    - **Operation:** Including linked sectors or clusters, it replicates all of the original drive's contents to help recover corrupted or erased information.

    - **Tools:** Disk-to-image files are often read using software programs like EnCase, X-Ways, FTK, and I Look Investigator.

2. **Bit-Stream Disk-to-Disk**

    - **Purpose:** Used when making disk-to-image files isn't possible because to mistakes or compatibility issues.

    - **Operation:** Bit stream copies are made using programs like Norton Ghost, EnCase, and Safe Back; these programs may also modify the characteristics of the destination disc so that it is identical to the source drive.

3. **Logical Acquisition**

    - **Purpose:** This strategy is designed to gather just the files that are necessary for a given case inquiry.

- **Operation:** In email investigations, investigators selectively retrieve just the.ost or.pst files from Outlook, or certain data from a massive RAID server.

4. **Sparse Acquisition**

   - **Purpose:** Used when scanning the whole disc is unnecessary, such as when only small portions of free space on the drive are relevant.

   - **Operation:** The investigators may gather and examine data bits that do not fit into the allotted storage space using this technique.

The versatility of these data gathering techniques allows for the preservation and copying of digital evidence with the assurance of data integrity and relevance to the case at hand, making them invaluable in forensic investigations and other domains. Considerations such as the kind of inquiry and the state of the initial data source dictate the analysis approach.

2.2 Techniques for Data Extraction

Data extraction is the process of gathering information from numerous sources and transforming it into a structured format that can be used for analysis or storage. This might be from databases, web pages, or papers. This process is essential in data integration workflows, enabling organizations to consolidate disparate data for informed decision-making. Effective data extraction streamlines workflows and enhances the efficiency of data management processes.

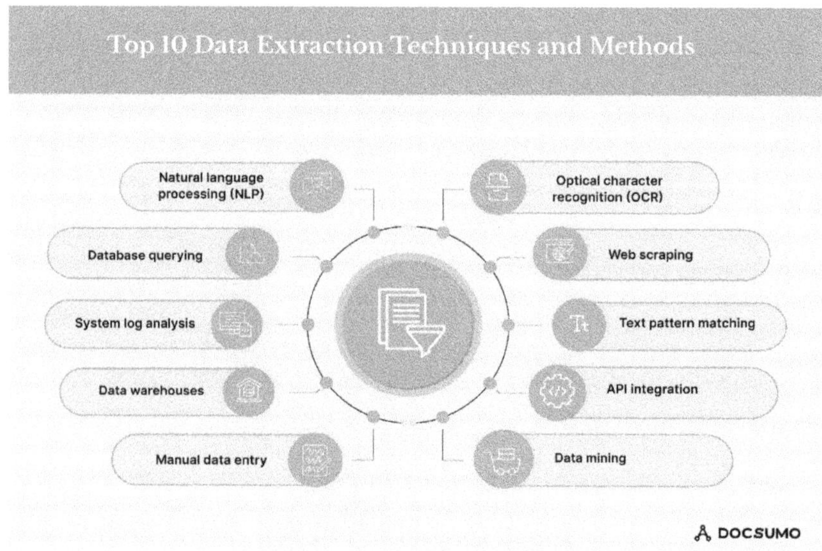

Figure 2.5: Data extraction

Source: - *(Ritu John, 2024)*

1. Web scraping

Scraping websites for their data is known as web scraping. It entails extracting data from websites (including text, photos, and links) and storing it in a structured manner for future use or study via the use of scripts or software.

Next, the data is transformed into a format that is compatible with spreadsheets and APIs.L

While you can perform web scraping manually, the term typically refers to automated processes by bots or web crawlers. It is a method of gathering and copying specific data from the web, which is then stored in a centralized local database for later retrieval or analysis.

Individuals and businesses can use scraping tools or libraries instead of manually noting down key data points (name, address, prices, dates, etc.)

How does web scraping work?

1. Using a web scraping tool, send an HTTP request to the target website's dedicated server to retrieve the HTML content of the web pages.

2. After a website grants access to the scraper, the HTML markup is parsed to identify and extract the required data elements. Parsing entails understanding the structure and arrangement of the HTML document and identifying particular HTML tags, attributes, or CSS selectors that are linked to the desired data.

3. The extracted and cleaned data is stored in a structured format like CSV, JSON, or a database for future reference and further analysis.

2. API integration

Quickly and easily access massive volumes of data from several sources using an API connection. It acts as a go-between for several systems, easing data flow and making it easier to get data from various sources (e.g., databases, internet, and software applications) without requiring individual access to each source.

An API allows for data centralisation by creating a single platform for all data and applications. The consolidation of these processes makes it possible to clean, prepare, and transport data without any hitches to its ultimate destination, such a data warehouse.

Banking, logistics, and insurance companies use OCR APIs to extract data from financial statements, invoices, and claims documents. Using a web scraping API to integrate the previously discussed data harvesting techniques with whatever app or project a particular business wants to implement is possible. This level of flexibility and adaptability is a crucial selling point of APIs in general.

How does API integration work?

1. After authenticating the user's identification with an API key, use the API documentation or instruction manual to make API calls to retrieve the desired data.

2. Once the API returns the data, you parse and extract the relevant information from the response. Data transformation may be necessary to get the information into a consistent structure that can be analysed or stored in your system.

3. The extracted data can be integrated into your analytics platform, business intelligence tools, or data warehouse. You can combine it with data from other sources to perform comprehensive analysis, generate insights, and create reports or visualizations

3. Text pattern matching

Text pattern matching is one of the data extractions approaches that involves searching a document or text for certain patterns or sequences of letters. It entails looking for regular expressions or predetermined patterns that fit a certain structure, format, or string of characters.

Data validation and pattern searches inside documents or large collections of documents are both made possible with this tool.

Pattern matching methods may be as basic as string matching and regular expressions for applications in NLP like grammar and voice recognition, or as sophisticated as ML algorithms for applications in financial analysis and fraud detection.

How does text pattern match work?

1. First, you define the pattern you want to match. It can be regular expressions, keywords, phrases, or other pattern definitions to determine the sequence you wish to search for.

2. Provide the text or document where you want to search for the pattern. It can be a paragraph, a document, or even an extensive collection of documents.

3. The text pattern matching algorithm processes the pattern and text input to identify matches. The algorithm typically scans the text input character by character, comparing it with the pattern to identify matches.

4. Depending on the requirements, the algorithm may iterate through the text input multiple times to find all possible matches.

4. Optical character recognition (OCR)

The electronic technique of making pictures with printed, typed, or handwritten text machine-readable is called optical character recognition (OCR). It can be performed on various sources, including scanned physical documents or digital images.

OCR solutions are essential to the banking, healthcare, and logistics sectors for automating data input, digitising documents, processing loan applications, bank statements, invoices, and receipts, and more.

How does OCR work?

1. The OCR tool acquires an image by scanning physical documents, files or websites.

2. The acquired image is preprocessed to enhance its quality and optimize it for processing. Techniques involve deskewing, DE speckling, script recognition, and various other adjustments.

3. It analyzes the preprocessed image and identifies individual characters or symbols using pattern matching or feature recognition. It matches the patterns and shapes in the image against a database of known characters.

4. After extraction, the text data is outputted in a digital format, such as PDF or word-processing document.

5. Data mining

Data mining is the process of discovering patterns in large datasets via the use of statistical analysis, machine learning, and database management systems.

The ability to make well-informed decisions, see trends, and forecast future outcomes are all made possible by it. Companies use data mining to find trends in consumer behaviour and use that information to improve their goods and services.

Similarly, financial institutions employ data mining to analyze credit card transactions and detect fraudulent activity.

How does data mining work?

1. The data that has to be mined, how it will be stored, and what format it will be presented in are all part of the first stage of data mining.

2. After selecting the information, the next steps are to clean, aggregate, and format it. An essential part of the data mining process, data transformation affects the efficiency and results.

3. After selecting and assessing suitable models, the next stage is to use data mining techniques on the dataset in order to reveal hidden patterns, correlations, and trends. It involves discovering associations, predicting outcomes, identifying anomalies, or segmenting the data into meaningful groups.

6. Natural language processing (NLP)

The field of NLP blends linguistics, computer science, and AI to investigate how computers and human language interact with one another. Its main goal is to efficiently handle and analyse large amounts of data that is defined by natural language.

Capturing the contextual intricacies and nuances inherent in language is the ultimate objective of making computers understand the content of papers. NLP technology is able to do this by correctly classifying and organising data, which allows it to extract useful insights.

There are several applications for NLP technologies such as chatbots, email filters, smart assistants, language translation, etc., ranging from analysing social media sentiment to communicating with clients.

How does NLP work?

1. The initial step involves preparing the text for analysis. It may include tasks like tokenization (breaking text into individual words or sentences), removing punctuation, converting text to lowercase, and handling special characters.

2. The next stage is called stemming or lemmatization, where the words are reduced to their root forms.

3. In the part-of-speech tagging stage, NLP assigns grammatical tags to words in a sentence, such as nouns, verbs, adjectives, or adverbs, to understand each word's role and syntactic context.

4. NLP methods detect and extract named entities (e.g., people, places, businesses, etc.) from text at the named entity recognition (NER) stage.

5. The next stage is Semantic analysis which focuses on understanding the meaning of words and sentences. It incorporates entity linking, sentiment analysis, semantic role labelling, and word meaning disambiguation. Semantic analysis helps interpret the text's intended meaning, sentiment, and contextual nuances.

7. Database querying

The process of pulling desired data or information from a database is known as database querying. The process entails communicating with a DBMS using the structured query language, SQL, in order to get data that meets certain criteria or constraints.

How does database querying work?

1. Deciding what information, you need to get is the starting point for defining the query. It includes specifying the tables and columns and any conditions or filters to narrow down the results.

2. Once formulated, the query is written in the appropriate syntax of the chosen database query language, such as SQL.

3. After writing the query, it is executed or run against the database. The DBMS processes the query and retrieves the requested data based on the specified criteria.

4. Once the query is executed, the DBMS returns the result set, which is the data that matches the query criteria. It is possible to aggregate, sort, filter, or do further analyses on the result set.

8. System log analysis

The log analysis method reviews, extracts, and interprets systems-generated logs. It can be done manually or using log analysis tools.

This is one of data extraction methods, that uses various techniques, such as pattern recognition, normalization, anomaly detection, root cause analysis, performance analysis, and semantic log analysis.

Log analysis helps improve security by detecting threats and cyber-attacks and mitigating associated risks.

9. Data warehouses

Document management systems known as data warehouses compile information from several sources and save it in one place for further analysis.

With the statistical analysis and data mining, visualization, and reporting features of a data warehouse, analysts and scientists can analyze historical records to derive insights that streamline business decision-making.

10. Manual data entry

Manual data entry is the process of manually employing data operators to input data into computer systems or databases. Businesses have been using this traditional data processing method for years. However, problems with manual data entry include increased errors and training costs.

These unavoidable risks with manual document processing have led businesses to adopt technologies to automate data extraction and achieve greater efficiency and accuracy.

2.2.1 Data Extraction Tools in 2024

Let's evaluate the best data extraction tools in 2024 with their features, pricing, and reviews. These tools use innovative data extraction methods to accurately and efficiently improve workflow automation.

1. Docsumo

Docsumo is an AI-powered data extraction platform that helps enterprises extract data from structured, semi-structured, and unstructured documents.

Pre-trained API models in Docsumo automatically capture key-value pairs, checkboxes, and line items from documents, freeing up employees for strategic tasks.

Businesses can also train custom models to extract data according to their requirements. The platform also helps with data analysis by studying trends and patterns and deriving actionable insights that optimize business operations and spending.

Key features

- Ingests documents from emails, document management systems, and scanners using API

- Extracts data accurately from documents using pre-trained API models

- Validates the data across original documents and against internal databases

- Routes the extracted data to relevant stakeholders for approval

- Integrates the extracted data with ERPs, CRMs, accounting, and payroll software solutions

2. Google Cloud Platform

Google Cloud's Document AI helps businesses build document processors to automate data extraction from structured and unstructured documents. This generative AI-powered platform effectively classifies documents and extracts crucial data from PDFs, printed texts, and images of scanned documents in 200+ languages.

Additionally, it offers advanced features such as recognizing handwritten texts (50 languages) and math formulas, detecting font-style information, and extracting checkboxes and radio buttons.

Key features

- Build custom processors to classify, split, and extract data from documents automatically

- Capture fields and values from forms using Form Parser

- Use pre-trained models to extract data from common document types such as paystubs, bank statements, invoices, US driver's licenses, and passport

3. Microsoft Azure

Microsoft Azure employs machine learning algorithms to extract texts, key-value pairs, and tables from documents. The platform has pre-trained models that automatically capture vital data from common documents such as receipts, purchase orders, and invoices.

The custom extraction capabilities let businesses extract tailored, reliable, and accurate data from documents.

Key features

- Pull and organize data from documents automatically without any manual labeling

- Customize data extraction results tailored to the layouts and templates

- Integrate with Azure's AI-applied search and easily find specific data in the documents

4. Amazon Web Services (AWS)

AWS Intelligent Document Processing powered by generative AI processes unstructured data files, classifies documents, captures vital information, and validates data against databases.

AWS IDP helps healthcare, insurance, legal, public sector, and lending industries automate their respective document processing workflows and improve efficiency.

Additionally, the platform automatically detects discrepancies such as missing digits in phone numbers, missing files, and incomplete addresses to ensure maximum accuracy.

Key features

- Ready-to-use APIs extract unstructured data from documents
- Synthesize information from multiple documents and get a quick summary to understand the context

5. IBM Watson

IBM Watson's Discovery lets businesses create models to process multiple documents and reports, capture data, and derive insights from them. The platform effectively discovers patterns and trends, helping businesses uncover hidden insights that can transform their specific business requirements.

IBM Watson's best features

- Smart Document Understanding (SDU) labels texts based on components such as headers, tables, and more
- NLP algorithms extract accurate information from all types of documents
- Users can train entity recognition models to extract specific entity types

2.3 Data Ingestion Frameworks

Data ingestion refers to the act of bringing information into a system from outside sources. Among the many data analytics process stages, it is essential. Many systems, such as those for financial management, CRM, social media, and email marketing, must be integrated into a company's data pipeline.

Data intake is usually done by data scientists since it calls for their knowledge of ML and programming languages such as R and Python.

Data Ingestion vs. ETL

There is a huge difference between data intake and ETL. Importing data into a storage engine or database is known as data ingestion, while ETL stands for extracting, transforming, and loading.

Due to their same meaning and frequent occurrence, distinguishing between the two might be difficult.

The primary distinction between ETL and data intake is the following:

- **Data Ingestion**

  The term "data ingestion" refers to the act of transferring information from one storage medium (such as a database) to another. When this is the case, the data is usually left unchanged.

  To illustrate the requirement of data ingestion, consider the following scenario: you have files stored in an Amazon S3 bucket that must be sent to your database.

- **ETL**

  The acronym "ETL" refers to the process of extracting, converting, and loading data from one system into another during data migration.

  This is in contrast to the usual practice of simply moving data from one place to another.

Data Ingestion vs. Data Integration

The processes of data intake and integration define the transfer of data between different systems. The first step is data intake, which involves adding information to a database; the second is data integration, which involves transferring that data from one system to another.

Data integration is often required when combining internal business operations with those of an external organisation or when using a product from one firm with another.

The two words' definitions vary from one another:

1. **Data Ingestion** - Data entry is the act of adding information to a database or other storage system. To transfer data from one repository, such as Salesforce, to another, such as SQL Server or Oracle, an ETL (extract, transform, load) tool is often used.

2. **Data Integration** - Data marging is the practice of integrating many databases into a single, usable model for usage by applications. This is especially common when the databases belong to separate companies, as is the case with Salesforce and Microsoft Dynamics CRM.

Types of Data Ingestion

The process of ingesting data into a data warehouse involves gathering information from several sources and cleaning it up. In order to analyse data, it must first be collected, cleaned, transformed, and integrated from several sources.

Two primary forms of data ingestion exist:

With real-time ingestion, data is continuously being sent into a data warehouse, often via cloud-based technologies that can swiftly receive the data, store it, and then make it available to users.

The process of batch ingestion is bringing together massive volumes of raw data from several sources in one location for automated processing.

Use this ingestion type when you need to arrange a lot of data before processing it all at once.

Benefits of Data Ingestion

Data intake is an essential component of any big data endeavour. The data entry procedure into your Hadoop cluster is known as "data pipeline" and it's not easy. Still, digesting your data has several advantages, such as:

- **Accuracy:** You'll be able to make sure that every piece of information you deal with is trustworthy and correct.

- **Flexibility:** You will find it much simpler to retrieve, alter, and analyse the data once you have consumed it, as opposed to utilising it in its raw form.

- **Speed:** Having all your data in one location can greatly accelerate processing times if you're utilising Hadoop for analytics or ML.

Data Ingestion Challenges

Data is a valuable resource. It's what allows us to be focused and productive, and it's the reason we can make choices. But how can you choose what data to retain and what to delete given the sheer volume of data?

Data quality, data collection, coding and maintenance, and latency are the four basic types of data input difficulties.

It may take some time to conquer the significant obstacles that are coding and maintenance. It could be more convenient to just delete old data than to spend time sorting it out for use in future endeavours.

When attempting to incorporate new data, businesses also confront the issue of latency. There could be major holdups in progress if you wait too long to consume your data and then use it in another program or procedure.

The lack of sufficient information or detail in older datasets is another problem with data quality. How often have you had to reprocess or clean up this data? It may take more than one pass through older files before they are suitable for our needs!

Lastly, there's the issue of gathering all this data initially—how can we start gathering all this data without erasing any of the necessary information?

Data Ingestion Tools

Any business can't function without data intake tools. Products in this software category collect and transmit data in a variety of formats, including structured, semi-structured, and unstructured. They streamline ingestion procedures that would otherwise need human intervention, freeing up time for businesses to make more informed choices based on data rather than transferring it.

The data intake pipeline is a set of interconnected processing stages that transfers data from one location to another. A data warehouse or reporting tool could be the next stop in the pipeline after an ETL tool has cleaned and formatted the raw data from a database or other source.

Data Ingestion Framework

A collection of services called the data ingestion framework (DIF) make it possible to load data into your database. Here are the parts that it contains:

Data may be retrieved from external sources, loaded into databases, or stored in Amazon S3 buckets for subsequent processing using the data source API.

The data source API proxy acts as a go-between for your app and the API. With this proxy in place, your app may communicate with other AWS services and resources, including Amazon S3 buckets, without you having to provide any additional authorisation information or credentials.

All the necessary code to communicate with external data sources via an API, in a manner analogous to web surfing (e.g., GET requests), is included in the data source service.

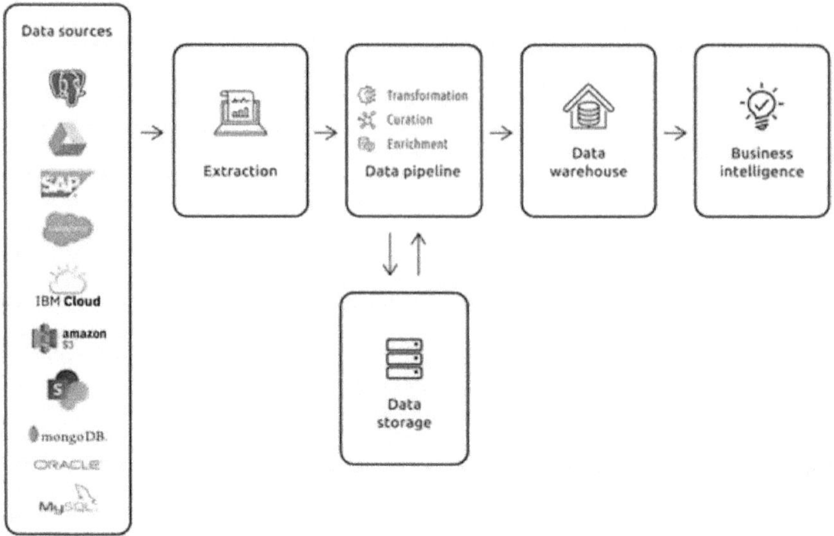

Figure 2.6: Data ingestion Framework

Source: - *(Ravindra Singh, 2022)*

Data Ingestion Best Practices

It may take some time and effort to build and implement a data pipeline properly. Additional data collection is necessary. Be careful to gather material in a format that your team can easily use in the future. Some guidelines for efficient data collection are as follows:

Gather only the information that is necessary at each step. Not having to reprocess anything later will save you time and money.

For easier data matching later on, it's a good idea to assign a timestamp or other unique identification to each item of information you gather. It will also aid in making sure your end product is accurate.

Ensure that all data is organised in a way that makes it easy for anybody to access and locate specific pieces of information in the future.

2.4 Relational Databases vs. NoSQL Databases

Relational Database: "Relational Database Management Systems" is the acronym for just that. One of the most well-known databases. Information is kept in it in the form of rows that are tuples. Because data is stored in tables, it has several tables from which data may be readily retrieved. E.F. Codd first suggested this model.

NoSQL: The term "non-SQL database" is short for just that. Unlike relational databases, NoSQL databases do not utilise tables to store data. For databases, it's the go-to method for storing and retrieving massive amounts of data. Better speed and support for query language are two of its benefits.

Difference between Relational database and NoSQL:

Relational Database	NoSQL
Data coming in at a low velocity may be handled by this.	Its purpose is to manage data that is arriving at a rapid pace.
Read scalability is all it provides.	Scalability in reading and writing is provided.
Structured data is managed by it.	It manages all type of data.
Data comes in from only a few places.	Data arrives from many locations.
Complex transactions are supported.	It supports simple transactions.
There is a single potential weak spot.	No single point of failure.
Less data volume is handled by it.	It handles data in high volume.

Relational Database	NoSQL
All transactions recorded in a single place.	Transactions written in many locations.
support ACID properties compliance	doesn't support ACID properties
Once a database is established, it is tough to make modifications to it.	Allows for quick and simple database updates
schema is mandatory to store the data	schema design is not required
Deployed in vertical fashion.	Deployed in Horizontal fashion.

Data Warehousing Concepts

Data warehouses are great for storing and analysing large amounts of information so that choices may be made with more precision. Data from many sources, including as relational databases, transactional systems, and others, is regularly fed into an organization's data warehouse (Inmon, 2002).

The goal of establishing a data warehouse is to improve decision-making via the storage, analysis, and interpretation of data. Data warehouses get frequent updates from a variety of sources, including transactional systems and relational databases.

Data warehouses are a kind of information management system that help with analysis and other business intelligence (BI) tasks. Data warehouses often hold massive volumes of historical data and are mainly intended to make searches and analysis easier.

The term "data warehouse" refers to a repository for an organization's collected information and data culled from both internal and external sources. Periodically, data is retrieved from a number of internal applications, including sales, marketing, and finance; applications that interact with customers; and systems maintained by external partners. After then, decision-makers are given access to this data so they may examine it. Data warehouse, then? To begin, it is a database that stores all sorts of information, both current and historical, with the purpose of improving a company's efficiency (Krishnan, 2013).

Key Characteristics of Data Warehouse

Here are some of the most important features of a data warehouse:

- **Subject-Oriented:** Data warehouses are subject-oriented because, instead of providing information about a company's general activities, they provide information about specific topics. Categories such as sales, promotions, inventories, etc. One must construct a sales-centric data warehouse, for instance, in order to study the company's sales statistics. A data warehouse like this would be a goldmine for answers to questions like "who was your best customer last year?" and "who do you think will be your best customer next year?"

- **Integrated:** The development of a data warehouse involves standardizing the format of data obtained from several sources. The name, formatting, and coding of the data stored in the warehouse must be consistent and adhere to global standards. This makes it easier to analyse data effectively.

- **Non-Volatile:** There can be no changes to data once it is placed into a data warehouse. The data is readable only. Entering new data does not delete existing data. The timing and nature of events may be better understood in this way.

- **Time-Variant:** Whether stated or not, the data kept in a data warehouse is accompanied by a temporal component. The Primary Key, which has to have a time component like a day, week, or month, is an illustration of how data warehouse time variation manifests itself.

Database vs. Data Warehouse

Data warehouses and conventional databases have several commonalities but are not necessarily synonymous. The key distinction is that data is stored in a database for various transactional reasons. On the other hand, analytics are performed by collecting data on a massive scale in a data warehouse. Data warehouses store information for use in large-scale analytical queries, whereas databases provide data in real-time.

An OLAP system, which answers database queries online, may be seen in data warehouses. Online database editing platforms, such as ATMs, are known as OLTP. Read on to find out how OLTP differs from OLAP.

Data Warehouse Architecture

There is often a three-tiered structure to data warehouse design.

- **Bottom Tier:** Data warehouse servers, which are typically located in the lowest tier, often stand in for relational database systems. Data cleansing, transformation, and feeding into this layer are all handled by back-end technologies.

- **Middle Tier:** There are two possible implementations of the OLAP server in the intermediate layer.

An enhanced relational database management system, the ROLAP paradigm converts multidimensional data processes into conventional relational ones.

Directly affecting data and actions with several dimensions is the MOLAP, or multidimensional OLAP.

- **Top Tier:** The data is extracted from the data warehouse using this front-end client interface. It has a number of tools, including those for data mining, analysis, reporting, and querying.

How Data Warehouse Works

Information and data gathered from several sources may be consolidated into a single, extensive database via data warehousing. Data warehouses often consolidate information from several sources, such as a company's website, mailing lists, point-of-sale systems, and customer feedback cards. Additionally, it may include sensitive personnel data, pay stubs, etc. Such data warehouse components are used by businesses for the purpose of consumer analysis.

Data mining is a data warehouse function that entails searching through large amounts of data for relevant patterns in order to come up with new methods to boost sales and profitability.

Types of Data Warehouse

There are three main types of data warehouse.

- **Enterprise Data Warehouse (EDW):** The enterprise-wide decision-support services are made possible by this form of warehouse, which acts as a central database. The benefits of this kind of warehouse include the ability to conduct complicated queries, a uniform approach to data representation, and access to information from across organisations.

- **Operational Data Store (ODS):** Data warehouses of this sort are updated in real-time. Commonplace tasks, like keeping personnel records, are typically better suited to it. This is necessary in cases when the reporting requirements of the company cannot be met by data warehouse solutions.

Data Mart

A data mart is a specialised section of a data warehouse that is created to support a certain division, area, or company function. There is a data mart or central repository for every department in a company. On a periodic basis, the ODS stores the data that comes from the data mart. Data is kept and used by the EDW after being sent to it by the ODS.

Data Warehouse Example

The data warehouse is an essential component of many businesses' day-to-day operations; let's examine a few instances.

The primary use case for data warehouses in the investment and insurance industries is the analysis of market and consumer trends, as well as related data patterns. Due to the massive losses that may occur from even a little discrepancy, data warehouses play an essential role in sub-sectors such as the foreign exchange and stock markets.

For marketing and distribution purposes, retail chains use data warehouses. This allows them to monitor products, review pricing strategies, and study consumer purchasing habits. In order to meet the demands of business intelligence and forecasting, they use data warehouse models.

However, data warehouse principles are used by healthcare organisations to provide treatment reports, communicate data with insurance companies, and conduct research and medical operations. In order to save lives, healthcare organisations rely significantly on business data warehouses to provide them with the most recent and accurate treatment information.

Data Warehousing Tools

Is the concept of data warehouse tools foreign to you? A large data collection may have several operations performed on it using these software components. These technologies are useful for retrieving, storing, and transferring data from a variety of sources. How may data warehouses be useful? Information processing tasks such as filtering, combining, sorting, etc. are their intended use (Ralph Kimball & Margy Ross, 2013).

Application types that use data warehouses include:

- Query and reporting tools
- Application Development tools

- Data mining tools

- OLAP tools

Oracle 12c, Xplenty, Amazon Redshift, Teradata, Informatica, IBM Infosphere, Cloudera, and Panoply are among the most well-known data warehouse solutions.

Benefits of Data Warehouse

Not sure why data storage is necessary for companies? Data warehouses provide several advantages for end users.

- Improved data consistency

- Better business decisions

- Easier access to enterprise data for end-users

- Better documentation of data

- Reduced computer costs and higher productivity

- Making it possible for end users to request reports or enquiries on the fly without negatively impacting the efficiency of operating systems

- Collection of related data from various sources into a place

Companies that have dedicated Data Warehouse teams outperform their competitors in crucial areas such as product creation, price, marketing, production time, historical analysis, forecasting, and customer happiness. Data warehouses will pay for themselves over time, despite their initial investment.

2.5 Cloud Storage Options

Google Drive (Windows, Mac, iOS, Android, Web)

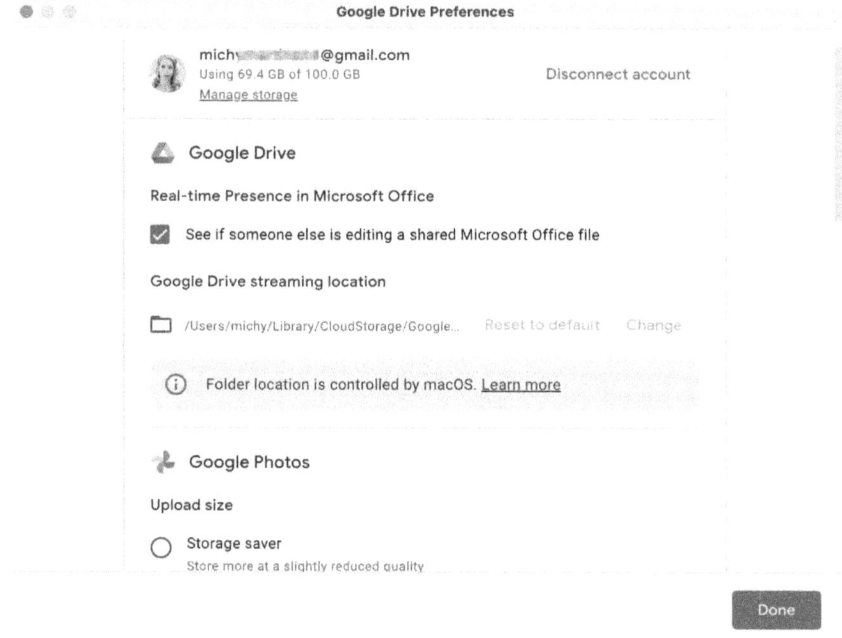

Google Drive pros:

- Built in to Android devices, including automatic photo backup

- Saves space on hard drives by storing data in the cloud, which can be accessed from any device.

- Powers storage of Google Docs, Sheets, Slides, and other Google Workspace files

Google Drive cons:

Forced to use it if you use Google Workspace

During the early 2000s, Google competed with the now-defunct Amazon Drive in the cloud storage market. Google Photos, which syncs with iOS and Android devices, is one of several automated backup features that

I've relied on for almost a decade now. Along with that, you should save any documents you create in Google Docs or any of the other Workspace applications to Google Drive.

Making the most efficient use of your device's hard drive capacity is a breeze with Google. Optionally, you have the choice to "stream" or "mirror" your files. Their streaming nature necessitates an active internet connection for their viewing. Because your files are "mirrored" on both your local machine and the cloud, you have access to them whenever you need them, regardless of whether you have an internet connection or not. I prefer streaming for huge folders that I don't use very frequently, and mirrored for files that I require quick access to at all times.

True story: I went to a coffee shop not long ago to get some work done, but they didn't have Wi-Fi. Fortunately, I had already mirrored my critical work files and enabled offline editing in Google Docs, so it was a non-issue. (However, amazing nevertheless.)

In comparison to iCloud, Google Drive's flexibility to link third-party applications is something I really like. As an example, I use Synology Cloud Sync to back up my actual NAS discs to Google Drive.

For my professional papers, I continue to use Google Drive since it is simple to use, required, and provides an easier (albeit not the most cost-efficient) option to back up my NAS discs. (I will get into this more at a later time.) Google Drive makes all the right moves for the majority of users.

Zapier makes it easy to link Google Drive with a wide variety of applications, giving you additional options. Todoist and Google Drive are my go-to tools for task management and file storage, respectively. With this Zap (automated workflow), whenever a new folder is created for a client project, it immediately adds a new task to my Todoist list.

Google Drive pricing: Get 15GB for free. Drive storage alone: $1.99/month for 100GB, $2.99/month for 200GB, $9.99/month for 2TB. Google Workspace: $6/month for 30GB, $12/month for 2TB, $18/month for 5TB.

iCloud (Mac, Windows, iOS, Web)

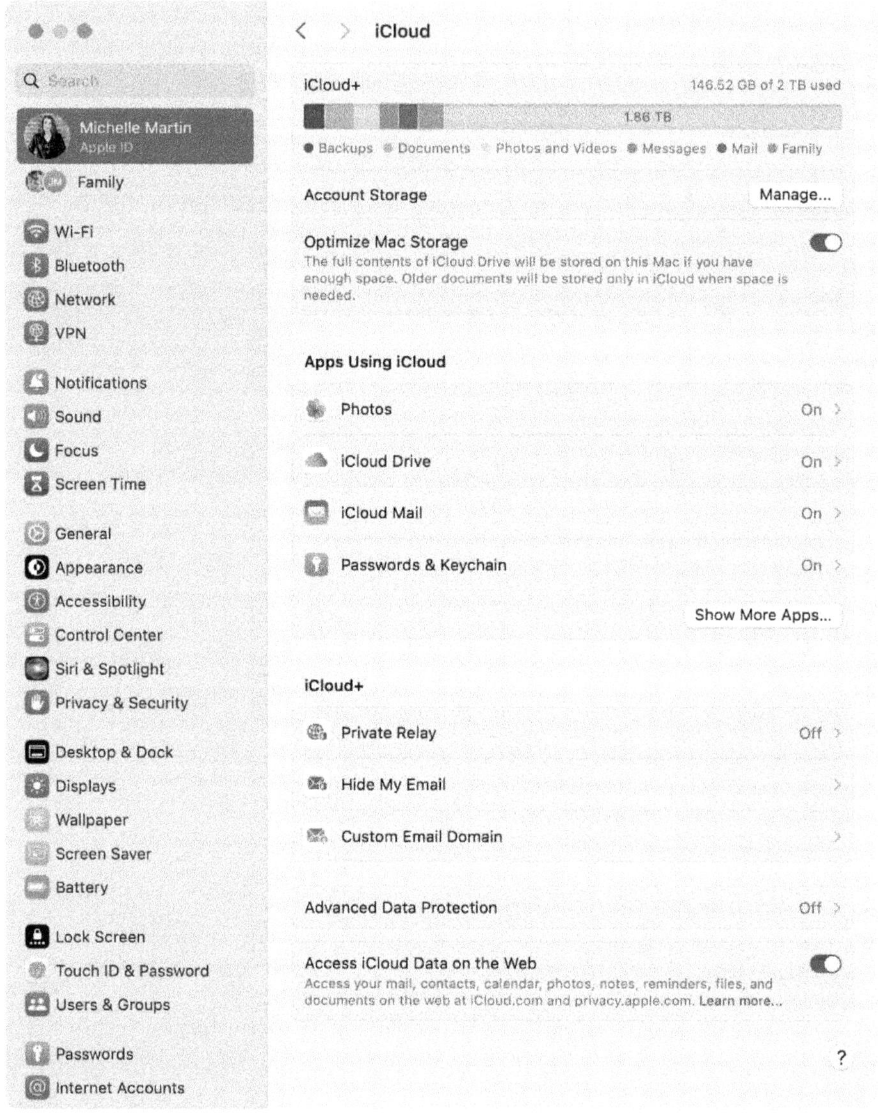

iCloud pros:

- Seamlessly integrates with all Apple devices
- Zero maintenance; it just works

iCloud cons:

- Photos take up the bulk of your storage space

- Not available cross-platform (e.g., Windows computer, Android phone, etc.)

I am you before you accuse me of being too critical of Apple loyalists. With my iPhone tucked into my pocket, I am penning this down on my MacBook, which sits next to my iPad.

The greatest advantage of iCloud, apart from its reasonable prices, is how well it integrates with Apple products. iCloud automatically backs up my iOS devices, including my phone, iPad, and my son's iPad. Oh my! I dropped my phone in the water! Got it—I've got all my things, even those expensive applications I've accumulated over the years.

Of course, iCloud backup also helps Apple out; it simplifies device upgrades to an absurd degree. When I updated all three of my Apple devices last year, the iCloud synced my applications, data, and settings in about an hour, so I could go straight back to how I liked it. Users love it and Apple makes money because of how convenient it is.

With iCloud, you don't need to install or update any applications; it's already a part of every Apple device. iCloud syncs flawlessly and never gives you any type of notice; it just runs in the background.

There are now 6TB and 12TB options available, making it one of the few cloud storage companies that offer bigger plans. There's a lot of space in the cloud. If you're using it to its full potential, a local storage solution, such as a NAS, will be more cost-effective. But iCloud is the way to go if you value simplicity above all else and want to continue purchasing Apple products.

iCloud pricing: Get 5GB for free. $0.99/month for 50GB, $2.99/month for 200GB, $9.99/month for 2TB, $29.99/month for 6TB, and $59.99/month for 12TB. Family sharing plans for up to 6 users are $3.99/month for 200GB and $12.99/month for 2TB.

Dropbox (Mac, Windows, iOS, Android, Web)

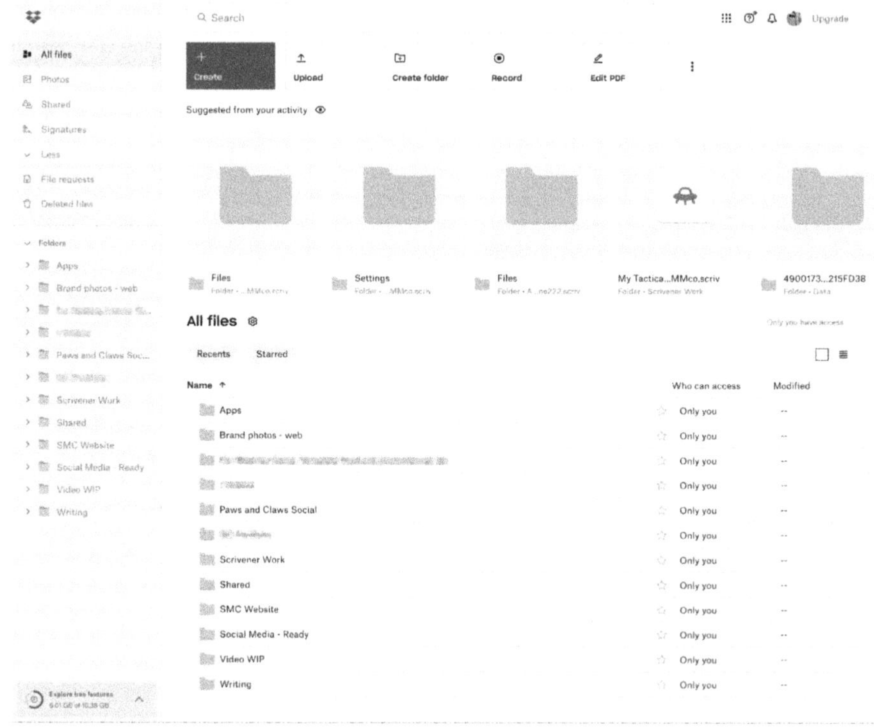

Dropbox pros:

- Low-maintenance app that just works
- Full cross-platform support

Dropbox cons:

- Removed their 2TB shareable family plan in 2024
- Not as competitively priced as other options on this list

Among the numerous competitive features offered by Dropbox's premium plans include the ability to remotely delete data in the event of device theft, an in-built password manager, and the ability to watermark images.

The fact that Dropbox is functional is, nevertheless, the most important factor, in my view. This stealthy program will quietly operate in the background on both Macs and Windows. For over a decade, I've relied on the free plan without a hitch, and that includes when it comes to synchronisation and offline file access.

Dropbox is also my go-to for file sharing. I sometimes get the sinking sensation of syncing—er, sharing—and wonder, "Am I sharing this file...or the entire folder?" even though Google Drive and others enable sharing. You can quickly identify which folders are shared and which aren't thanks to Dropbox's iconography.

You can share your yearly folder of tax trash with your accountant or save a lot of papers on it, even if the free storage limit is minimal. Dropbox is the only online storage solution officially supported by the popular writing program Scrivener, which is very specialised for people like me who write. Dropbox is the greatest supported sync option for Scrivener, and it's also the easiest cloud storage program to use if you value simplicity above everything else.

Zapier is compatible with Dropbox, so you can automate the saving of forms and email attachments, app updates, and a whole lot more. Although the number of possible Zaps is almost endless, my favourite is the one that allows me to save attachments from Notion databases to Dropbox. Learn more about the automated ways to organise your Dropbox or use one of these templates to get started.

Dropbox pricing: Get 2GB for free. $11.99/month for 3TB. Various business plans start at $54/month for 3 users and 5TB.

Box (Mac, Windows, iOS, Android, Web)

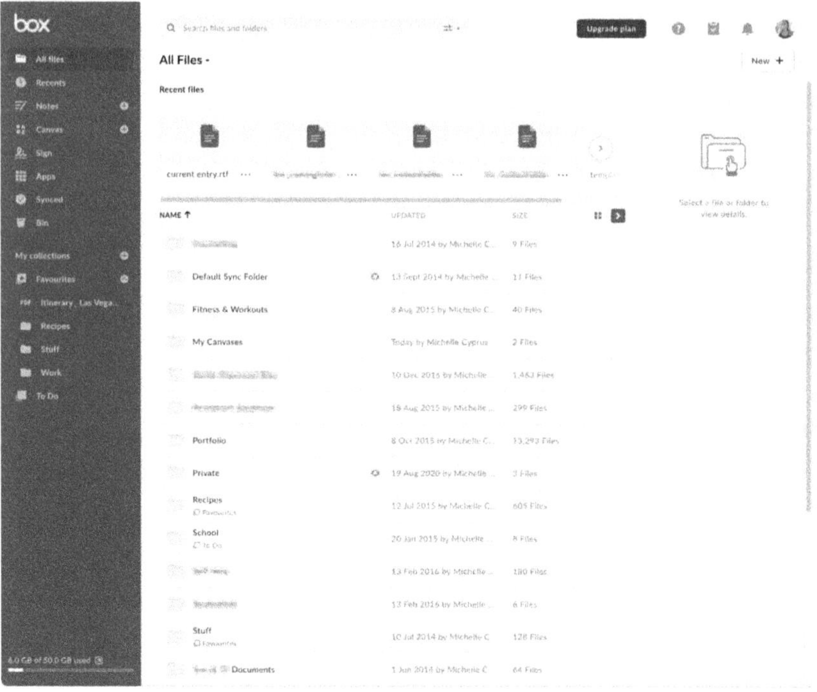

Box pros:

- Essentially unlimited storage for $60/month

- Document signing, online document collaboration, and whiteboarding are just a few of the small company capabilities included in this bundle.

Box cons:

- Unlimited storage is capped at 5GB per file

- Pricing only becomes competitive at higher storage tiers

However, Box does have many positive aspects. Its smartphone app is the most user-friendly, in my opinion. You can easily access important files or folders on mobile using the Favourites and Collections tabs. This feature works regardless of where you are, so it's perfect for accessing on the move.

The ease and speed with which one can edit Box Notes (text documents) on mobile makes it ideal as a content production centre for small enterprises or as a solution for taking notes while on the go that syncs across devices. Anywhere you are, working together is a snap thanks to the comment system in the file. If you're looking for a virtual whiteboard for individual or group brainstorming, Box has you covered with Canvas.

Assuming you don't have a need to keep a great deal of video or big, unique files, Box Sign's document signing capabilities and those of Box make it a formidable rival for small enterprises and freelancers.

Zapier is an app that works with Box, so you can automate things like sending files to your Box storage via email or updating your remote team when resources are added.

Box pricing: Get 10GB for free. Plans range from $14/month for 100GB to $60/month (minimum 3 users at $20/each) for unlimited storage with a business account.

OneDrive (Mac, Windows, iOS, Android, Web)

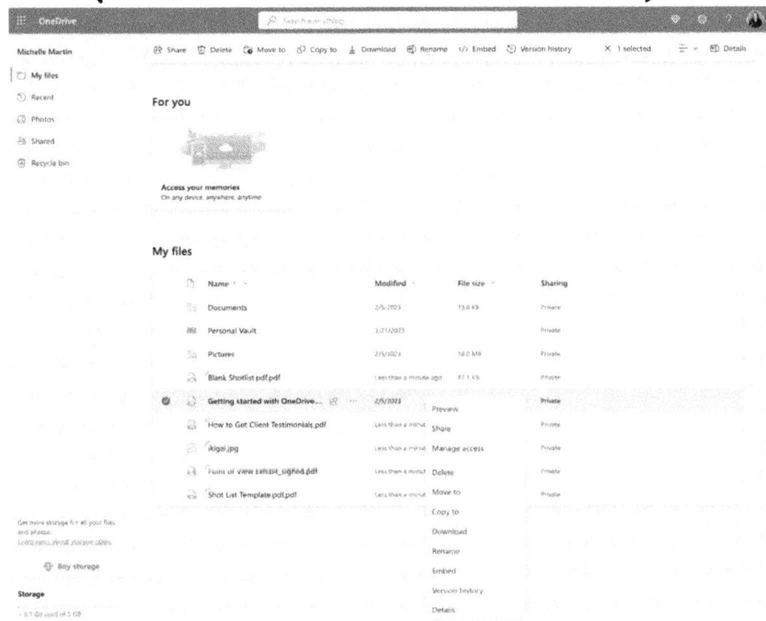

OneDrive pros:

- Built in to Microsoft/Windows devices but still cross-platform on macOS and iOS

- Organized photos library and galleries

OneDrive cons:

Storage space is quite limited; 1TB max per person on 6TB plan

As a user of Microsoft Office, you should really consider OneDrive so you can access and edit your Word, PowerPoint, and other files from any location. Not only does it function admirably on Mac and iOS, but its deep interaction with Windows further enhances its convenience. The online, desktop, and mobile applications are all structured in a logical way. Its price is also comparable to iCloud.

The revamped Photos tab, which sorts your images into folders according to things like selfies and screenshots and presents them in a visually beautiful gallery reminiscent of Apple Photos, may be even more crucial to some people.

So what's exciting about OneDrive? Not much.

Whiteboarding features and personalised branding are being introduced by competing programs, whereas OneDrive only serves to store your files. Apps that excel at only one thing aren't always the best. Plus, given Microsoft's stellar reputation, OneDrive handles storage well. This is obviously the best option for Windows and Office users. Because of how dependable and user-friendly it is, it ought to be a formidable competitor for everyone else.

In addition, the automations I described earlier can be done using OneDrive's integration with Zapier, and you can utilise Zapier to make your apps work better on other platforms. Would you like to work using Google Docs instead of Google Drive? Use OneDrive to keep them safe.

OneDrive pricing: Get 5GB for free. Paid plans are $1.99/month for 100GB, $6.99/month for 1TB, $9.99/month for 6TB (shared with 6 users, split 1TB each).

Jottacloud (Mac, Windows, iOS, Android, command line tool)

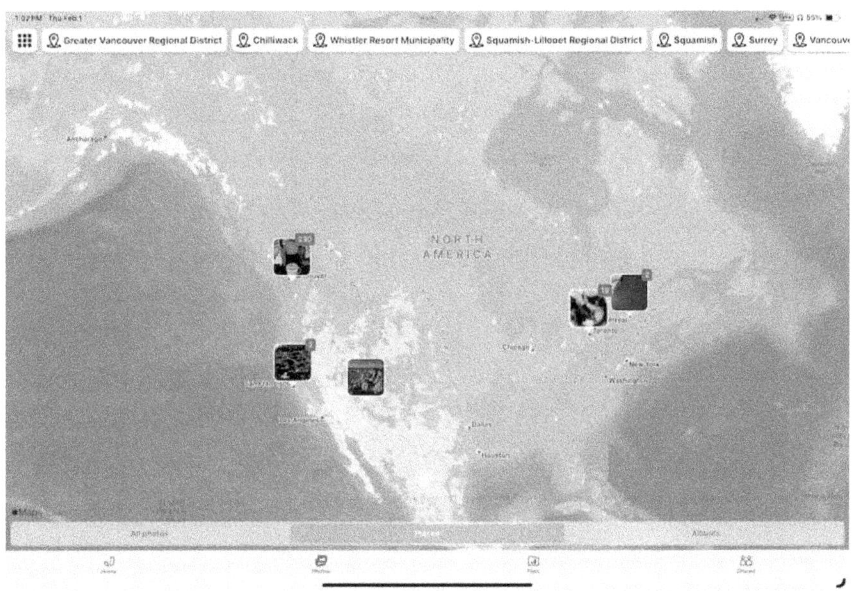

Jottacloud pros:

- Truly unlimited storage space
- Automatic photo backup on any device

Jottacloud cons:

Desktop app can be a bit buggy and crash sometimes on macOS

Jottacloud is great for organising media files, but it can handle any kind of file. Jottacloud is a mobile software that can back up your phone's images and videos automatically, much like Apple images and Google Photos. Immediate storage space may be made available on your device by erasing uploaded material with a single click. You can share anything, including RAW files, to streaming media players like Apple TV and Chromecast, and the app's gallery views of your memories are stunning.

Oh, did I mention unlimited storage is under $10 per month?

In order to safeguard my network attached storage (NAS), various mobile devices, phone photographs, and PC hard drive, I began using Jottacloud a year ago. It replaced Apple Photos, met most of my storage requirements for iCloud and Google Drive, and supplanted Back blaze, my long-standing PC backup solution. By combining all of those requirements into Jottacloud's inexpensive unlimited package, you may save money, and the picture organisation tools are fantastic.

According to Jottacloud, you may expect 5 terabytes of speedy storage, and after that, you'll have infinite storage with "reduced upload speeds." My anticipation of sadness at the prospect of giving up my lightning-fast backup speeds grew as I neared 5TB over the course of a year. Little did I expect, nothing has changed. It could be because my Wi-Fi is so sluggish, but Jottacloud is an incredible game-changer when it comes to limitless storage, especially for big media libraries and backups of hard drives.

With cloud storage, you can pick which files to sync with your device and which to keep in a one-way mirror. Even external hard drives may have their data automatically backed up via Jottacloud. Because Jottacloud is entirely owned by Norwegians, it is not subject to the U.S. CLOUD Act, thus your privacy is likewise guaranteed.

Jottacloud pricing: Get 5GB for free. Unlimited storage is €9.90/month for personal use. Family sharing plans (5 users) start at €6.90/month for 1TB and up, and business plans licensed for commercial use range from free (5GB) to €49/month for 1TB storage.

Koofr (Mac, Windows, Linux, iOS, Android, Huawei AppGallery, WebDAV, Rclone, Web)

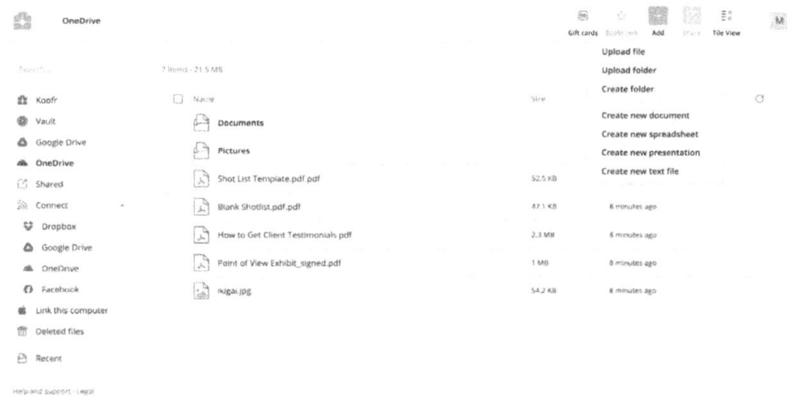

Koofr pros:

- Merge free accounts from different storage providers or even different companies into one.

- Only pay for the storage you need

Koofr cons:

- It takes some planning and administration, as opposed to a hands-off approach, to choose the appropriate storage add-ons for your requirements.

Straight up: Discover the perfect app for automatic Huawei phone backups with Koofr. A general-purpose cloud storage manager is where Koofr really shines for the remainder of you.

Koofr stands out from the competition since it integrates with popular cloud storage services like Google Drive, Dropbox, and OneDrive in addition to its own built-in storage (10GB for the free tier). From any device, you can access all of your cloud storage apps via the Koofr app, where you can manage files and do searches.

This is a fantastic method to consolidate all of the free plans from different applications into one neat package if you're on a tight budget like me. An important feature is that you may link several accounts to different third-party apps. Consider creating two distinct free Google Drive accounts, one for business data and one for personal stuff.

To accommodate situations where you may want greater transfer limitations for a single bucket but not for the others, Koofr's mix-and-match pricing plans let you add on "buckets" with varying storage capacities and features. Paid subscriptions also come with extra features including a PDF/image editor, unrestricted file sharing, online editing for Microsoft Office, and 30-day file recovery.

I admit: I was reluctant to consolidate all of my disparate cloud storage services into one. I was worried it would be cumbersome, but the robust cross-platform search really increases efficiency.

Koofr pricing: Get 10GB for free. Plans start from €0.50/month for 10GB to €120/month for 20TB, and you can mix and match storage blocks and features to suit your needs.

iDrive (Mac, Windows, Linux, Unix, iOS, Android, Windows Server)

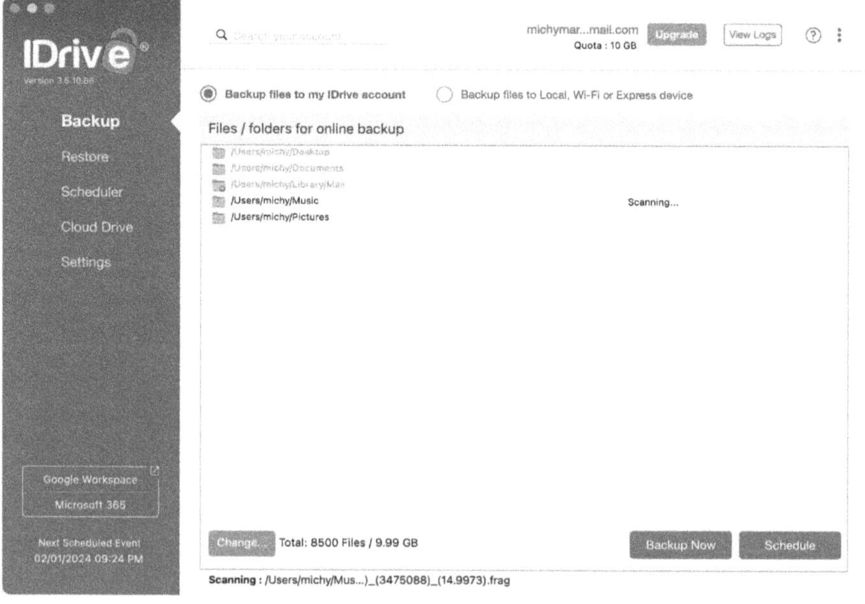

iDrive pros:

- Comprehensive backup solution, including physical drive shipments
- Affordable storage up to 100TB

iDrive cons:

Data privacy compliance only guaranteed with Team and Business plans

iDrive is an all-inclusive, reasonably priced cloud storage tool that works well for almost any use case: it can back up devices (including mobile phones and NAS drives) and PCs automatically, store anything in the cloud, sync data between devices, and much more.

The two terms "cloud storage" and "cloud backup" are not interchangeable. Storing data in the cloud is the same as moving individual folders and files to the cloud. In the case of a hard drive failure, you may easily recover a device with a single click thanks to cloud backup, which involves automatically backing up an entire device in the background, including cloning and versioning. Both can benefit from iDrive. Even better, iDrive will send you a physical copy of your backup in case your hard drive fails, allowing you to recover your device without using your data plan.

As soon as you connect an external hard drive or networked drive, such as a NAS storage system, to your computer, iDrive begins scanning all of the devices you've linked to it for changes and creates backups of those data.

You won't find a more insane quantity of cloud storage than iDrive's 100 TB/month plan, which costs around $1 per TB per month. However, at lesser levels, iDrive isn't the most inexpensive option.

For the convenience of backing up several PCs or hard drives, or just having access to a big amount of material from any location, iDrive's increased monthly fees may be justified.

iDrive pricing: Get 10GB for free. $9.95/month for 5TB, $14.95/month for 10TB, and $24.95/month for 20TB, $49.95/month for 50TB, and $99.95/month for 100TB.

Internxt (Mac, Windows, Linux, iOS, Android, Web)

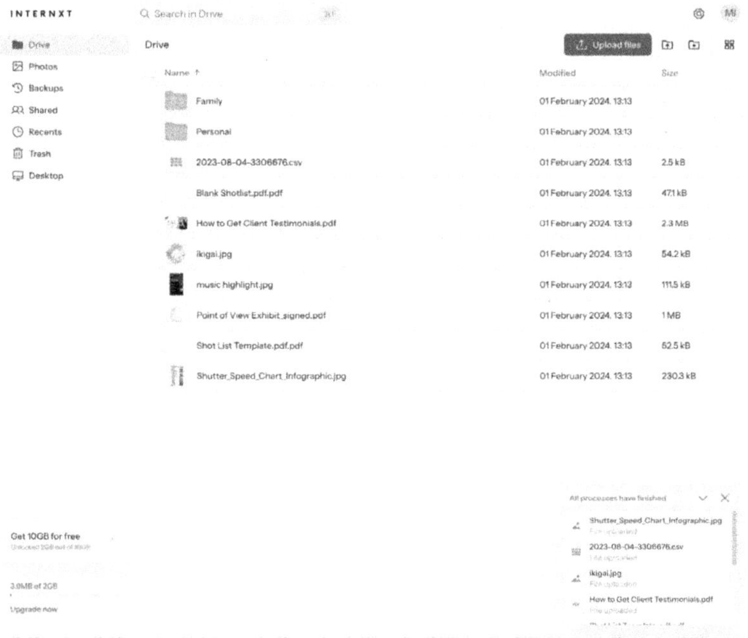

Internxt pros:

- Maximize data privacy without needing technical knowledge
- Separate area for photo backup, with gallery organization

Internxt cons:

Pricing per TB isn't as competitive as other options

Internxt isn't the only cloud storage program that prioritises privacy; Proton Drive also offers private email, calendaring, and a VPN. However, it achieves the ideal mix between data privacy and usability. By operating similarly to a "traditional" cloud storage service, Internxt streamlines security, while Proton and others need a little more technical know-how to work properly.

Internxt has state-of-the-art features including secure file sharing and transfers, zero-knowledge AES-256 encryption, and cross-platform applications that are simple to use immediately. The interface is extremely Dropbox-y. With military-grade encryption, your images are protected even when you back them up and access them from any device. There's a dedicated section for that.

Internxt may be the ideal cloud storage software for you if you agree that privacy should be considered a basic human right and not an optional extra.

Internxt pricing: Get 10GB for free. $5.49/month for 200GB, $10.99/month for 2TB, $22.99/month for 5TB, and $34.99/month for 10TB.

MEGA (Mac, Windows, Linux, iOS, Android, Chrome extension, Edge extension, Command Line Protocol, QNAP NAS, Synology NAS)

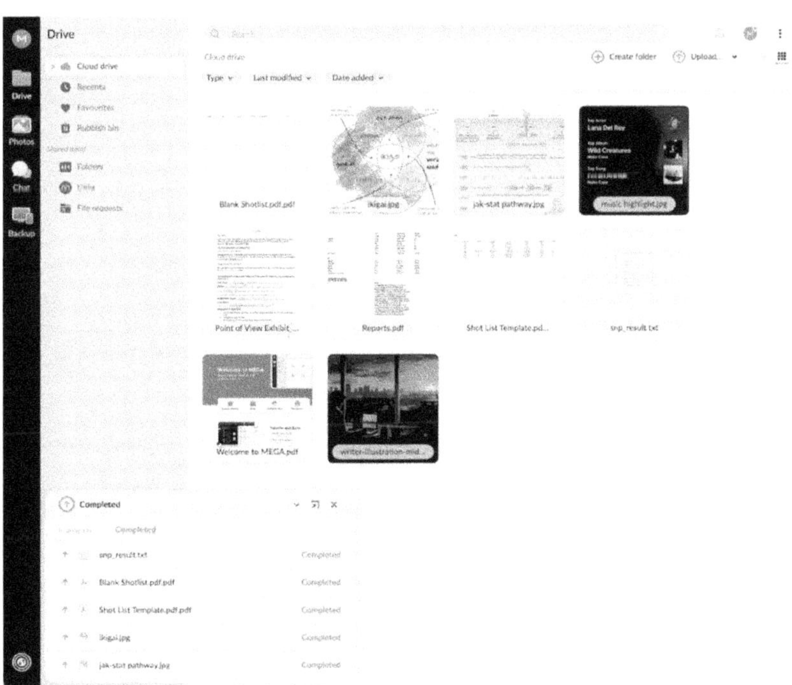

MEGA pros:

- Best value for money at 8TB level and up

- Officially supported NAS packages and command line tools for advanced users

MEGA cons:

Monthly file transfer limits

Lose your password = lose your account (due to privacy encryption—considered a pro for some)

At the 8 terabyte (TB) level and above, MEGA provides the most bang for the buck among competing providers. Individual plans are available up to 16 TB. Beyond its reasonable pricing, MEGA provides zero-knowledge encryption for maximum privacy and a plethora of options for techies, such as Synology and QNAP bundles for NAS live synchronisation, versioning, and scheduled backups.

In fact, scriptable commands may be executed straight from the command line in Windows, Linux, or Mac OS X, allowing super-geeks to automate and manage MEGA processes.

The only real drawback of MEGA is that you can only transmit files up to the amount of storage space your subscription allows each month. So, if your cloud storage capacity is 8 terabytes, you may transmit 8 terabytes of data every single month. Whether you're backing up devices to MEGA for the first time or often back up several devices and/or huge data, such as 4K video or RAW photographs, you're sure to get near to these generous limitations.

All around, it's hard to beat MEGA thanks to the price and advanced functionality.

MEGA pricing: Get 20GB for free. €9.99/month for 2TB, €19.99/month for 8TB, €29.99/month for 16TB (Martin, 2024).

Multiple Choice Questions (MCQs)

1. **Which of the following is NOT a typical data source for data engineering?**

    a.   Relational databases

    b.   Social media platforms

    c.   Data lakes

    d.   Spreadsheet documents

2. **What is the primary goal of data extraction in data engineering?**

    a.   To clean and filter raw data

    b.   To move data from one system to another

    c.   To transform the data into useful formats

    d.   To store the data for future use

3. **Which of the following is a common framework used for real-time data ingestion?**

    a.   Apache Kafka

    b.   MySQL

    c.   MongoDB

    d.   Hadoop HDFS

4. **Which of the following best describes a key difference between relational databases and NoSQL databases?**

    a.   NoSQL databases use structured data models, while relational databases use unstructured data models.

    b.   Relational databases are more flexible in handling large-scale, distributed data.

    c.   NoSQL databases are more suited for handling unstructured or semi-structured data.

d. Relational databases do not support transactions, while NoSQL databases do.

5. **What is the primary function of a data warehouse?**

a. To store data in a transactional format

b. To collect, store, and analyze large volumes of historical data

c. To store real-time data for operational purposes

d. To store unstructured data in raw format

6. **Which of the following is a popular cloud storage option for data engineering?**

a. Amazon S3

b. SQL Server

c. MongoDB

d. Oracle Database

7. **Which of the following data sources is most commonly used for extracting large volumes of real-time data from applications?**

a. Data lakes

b. APIs (Application Programming Interfaces)

c. Relational databases

d. Flat files

8. **Which method of data extraction involves pulling data in small batches at scheduled intervals?**

a. Real-time extraction

b. Batch extraction

c. Stream processing

d. On-demand extraction

9. **In which scenario would a NoSQL database be more suitable than a relational database?**

    a. When the data is structured and requires complex querying

    b. When there is a need for high scalability and flexible schema design

    c. When the database supports only a single table for storage

    d. When the database requires ACID transactions

10. **What is an example of a process involved in data warehousing?**

    a. Data cleaning and transforming

    b. Data replication in real-time

    c. Storing data in JSON format

    d. Continuous data streaming

Answer

1	2	3	4	5	6	7	8	9	10
D	B	A	C	B	A	B	B	B	A

Chapter 03

DATA PROCESSING AND TRANSFORMATION

3.1 Data Cleaning and Preprocessing

Data science and ML place a premium on high-quality data. A machine learning model's accuracy is highly dependent on the quality of the input data. Here, pre-processing and data cleansing are essential parts of the ML pipeline, not only first stages.

Errors in the dataset may be found and fixed during data cleaning. This includes issues like managing outliers, duplicates, and missing or inconsistent data. Make sure you train the ML mode on dependable and correct data. wrong data may be learnt by the model without adequate cleaning, which may result in wrong classifications or predictions (Kleppmann, 2017).

Data pre-processing, on the other hand, include data cleansing and other processes that get data ready for ML algorithms. Data reduction, normalisation, feature selection, and transformation are some of the possible processes in this process. Transforming raw data into a format that ML algorithms can understand is the main objective of data pre-processing. The process may be taken to the next level with the help of data science consulting services, which provide knowledge to optimise data transformations and get data sets ready for sophisticated analytics.

There is no way to exaggerate the significance of data pre-processing and data cleaning in determining the model's success. Machine learning models that are trained on datasets that have been properly cleaned and pre-processed tend to provide more accurate and trustworthy results, while datasets that have been improperly cleaned and pre-processed tend to produce misleading conclusions.

Data Cleaning

Data quality is king in the data science and ML realms. It's common knowledge that ML model performance is significantly affected by data quality. Because of this, finding and fixing (or erasing) incorrect or corrupt entries from a dataset is an essential part of data science.

Erasing or replacing data is just part of data cleansing. To prepare raw data for analysis, it is necessary to undergo a thorough procedure including several approaches. Data type conversion, addressing missing values, and duplicate removal are all part of these methods. The kind of data and the needs of the analysis dictate the use of the appropriate approach.

Common Data Cleaning Techniques

1. **Handling Missing Values:** There are a number of potential causes of missing data, including mistakes made during data collection or transmission. Various approaches may be used to deal with missing data, which is dictated by the kind and quantity of the missing information.

    - **Imputation:** Substituting values for missing ones is the goal here. The replaced value can be the most common category in a set of categories or a central tendency measure like the mean, median, or mode in numerical data. Two examples of more complex imputation techniques are multiple imputation and regression imputation.

- **Deletion:** You clean up the dataset by removing the cases when values are absent. Despite its simplicity, this strategy may result in data loss, particularly in cases when the missing data is not random.

2. **Removing Duplicates:** Data merging and data input mistakes are two common causes of duplicate entries. The data might be distorted and the findings can be skewed due to these duplications. Finding these repetitive items using important characteristics and deleting them from the dataset is one way to remove duplicates.

3. **Data Type Conversion:** In some cases, the data may not be in the right format to be used with a certain model or analysis. A number attribute, for example, might be stored as a string. When this happens, we alter the data type of an attribute or collection of attributes by using data type conversion, which is also called datacasting. Preparing the data for processing by machine learning algorithms is the first step in this procedure.

4. **Outlier Detection:** Data points that stand out from the rest are called outliers. They may arise as a result of mistakes or data variability. In order to find these outliers, outlier detection methods are used. Statistical methods like the Z-score and IQR as well as machine learning algorithms for anomaly identification and grouping are examples of these approaches.

The data science pipeline would not be complete without first cleansing the data. Machine learning models are strengthened and made more trustworthy when accurate, consistent, and reliable data is used for analysis and modelling.

Steps to Perform Data Cleanliness

Data cleaning is the process of systematically finding and fixing mistakes, discrepancies, and inaccurate information in a dataset. In order to clear up your data, you must follow these procedures.

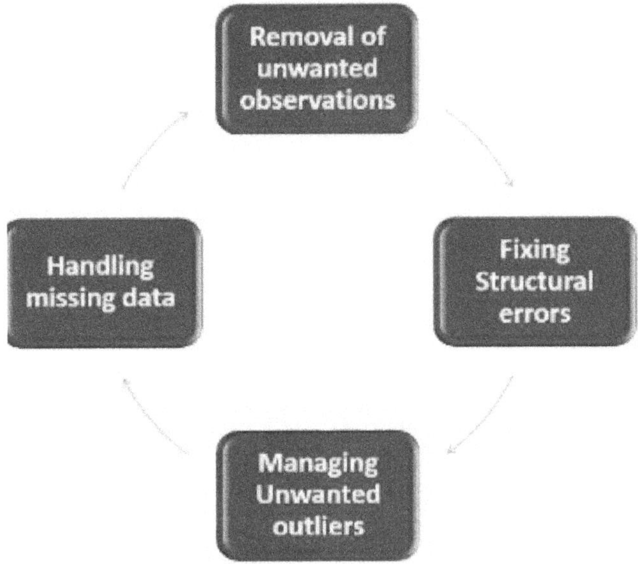

Figure 3.1: Perform Data Cleanliness

Sources: *- (Geeksforgeeks, 2024)*

- **Removal of Unwanted Observations:** Sort the dataset by relevance and remove duplicate or irrelevant observations. During this stage, all data entries are reviewed for duplicates, irrelevant details, or data points that do not add value to the study. Datasets may be made more efficient by removing superfluous observations; this also lowers noise levels and improves overall quality.

- **Fixing Structure errors:** Take care of the underlying problems with the dataset, including inconsistent data formats, name standards, or variable types. Make sure all data is represented consistently, standardise formats, and fix name conflicts. Correcting structural mistakes improves data consistency and makes it easier to analyse and understand accurately.

- **Managing Unwanted outliers:** Locate and handle data points that are very out of the ordinary, known as outliers. Determine whether outliers should be removed or transformed to reduce their influence on the study, taking into account the context. If you want more trustworthy insights from your data, managing outliers is a must.

- **Handling Missing Data:** Methods for efficiently dealing with missing data should be developed. Statistical approaches, complex imputation techniques, or the removal of records with missing data may be required to do this. A more comprehensive dataset, free of biases, is the result of properly handling missing data. This guarantees that studies are conducted with integrity.

Best Practices for Data Cleaning

Here are some practical tips and best practices for data cleaning:

- Importing new data requires stringent adherence to data quality standards.

- Fix errors and fill in missing sections using efficient and precise algorithms.

- Using known criteria and cross-checks, validate the data.

- Data cleansing is an ongoing activity, not a one-and-done deal. It is important to clean and pre-process incoming data as it arrives before adding it to the model.

Following these guidelines will help us maintain clean and organised data, which will in turn improve the efficiency of our machine-learning models.

Data Pre-processing

Data pre-processing is an essential element of data science, especially when it comes to using machine learning. In order to make the dataset more acceptable for machine learning algorithms, it must be prepared and cleaned. This method may make the model easier to understand, less likely to overfit, and more effective in general.

Exploratory data analysis is the first step in data pre-processing as it helps you understand the intricacies of your dataset and the primary problems with the data. There are a lot of mistakes, missing data, inconsistent scaling, and other issues with real-world data. You need to fix these things so people can utilise and comprehend the data. The data pre-processing procedure involves cleansing and fixing the majority of the data errors.

The efficiency of your ML model and subsequent operations could be compromised if you neglect the data pre-processing stage. While certain models may be able to deal with outliers, high dimensionality, and noisy data, the vast majority of models struggle with missing values. In order to make the required corrections to the data before feeding it into your ML model, pre-processing is essential. This is because it completes and improves the dataset.

Data cleansing, dimensionality reduction, feature engineering, data sampling, data transformation, and unbalanced data management are all part of the pre-processing procedures. The data problems that each of these approaches aims to solve are unique to that strategy.

Common Data Pre-processing Techniques

1. **Data Scaling:** The range of data's independent variables or characteristics may be standardised via data scaling. In order to avoid one characteristic overwhelming the others, particularly with big datasets, it attempts to standardise the data's range of features. For data-range-sensitive algorithms like deep learning models, this is an essential aspect of data preparation.

   Min-Max normalisation and standardisation are two of the many methods for scaling data. Data is often scaled from 0 to 1 using Min-Max normalisation, while data is scaled with a mean of 0 and a standard deviation of 1 using Standardisation.

2. **Encoding Categorical Variables:** Numerical inputs are necessary for machine learning models. Encoding categorical input to numerical values is a prerequisite to model fitting and evaluation. An often-used method for pre-processing data, this is called encoding categorical variables. One-Hot Encoding is a popular technique; it takes the original columns and generates new binary columns for each label or category.

3. **Data Splitting:** A common data splitting strategy is to create separate sets for training, validation, and testing. To fine-tune the model's parameters, you use the validation set after training it with the training set. In order to evaluate the final model objectively, the test set is used. To avoid having the model get overfit to a small sample of the data, this method is crucial when working with big datasets.

4. **Handling Missing Values:** Results could be skewed if certain data points are missing from the dataset. That is why it is critical to deal with missing values properly. One way to deal with missing data is to remove the rows that have them. Another is to imputation, which involves replacing the missing values with statistical measurements like as the mean, median, or model. To guarantee high-quality data for training ML models, this step is essential.

5. **Feature Selection:** Feature selection is a machine learning method that involves automatically choosing data attributes that have the greatest impact on the prediction variable or output that you're targeting. Many models, particularly linear methods such as logistic and linear regression, may be inaccurate when data contains irrelevant elements. Data scientists dealing with high-dimensional data should prioritise this procedure since it decreases training time, increases accuracy, and lowers overfitting.

Three benefits of performing feature selection before modeling your data are:

- **Reduces Overfitting:** Having fewer duplicate data reduces the chances of making conclusions based on noise.

- **Improves Accuracy:** The accuracy of the models is enhanced when there is less misleading data.

- **Reduces Training Time:** The algorithm becomes simpler with fewer data points, and it trains more quickly.

Importance of Data Cleaning and Data Pre-processing

An essential part of doing valid analyses is cleaning and preparing the data so that the results are accurate and reliable enough to inform decision-making. Decisions made without following the correct procedures could result in erroneous and unreliable insights. Data that is either inaccurate or incomplete could cause overestimations or underestimates, making it harder to spot important patterns and trends that are vital to making good decisions.

- **Data pre-processing and data cleaning** mitigate the risks of data bias. A bias exists when the results of a study are skewed towards a given group, whether that group benefits or suffers from the data used to draw those conclusions. Incomplete data, erroneous representation, and insufficient sample methods may all lead to bias. In order to lessen or remove data bias, cleaning and pre-processing are necessary steps.

- **Data cleaning and pre-processing** eliminate outliers, which impact data analysis. When compared to the rest of the numbers in the dataset, outliers stand out as very unusual. The statistical findings of the study may be impacted by them since they have the potential to significantly alter the statistical measurement results. Organisations depend on predictive models, but erroneous findings might compromise their reliability.

- **Data pre-processing and data cleaning** promote better data organisation, which can help with the simplicity and efficiency of the analysis. It improves the analysis's accuracy while also making data protection and presentation as simple and straightforward as feasible. Data that is well-organised helps to cut down on analysis time and resources, leading to overall efficiency gains.

Differences Between Data Cleaning and Data Pre-processing

Aspect	Data Cleaning	Data Pre-processing
Scope of Activities	Problems, discrepancies, and missing values must be located and fixed. Focuses on ensuring data accuracy and completeness.	Data cleansing, feature scaling, feature selection, transformation, encoding, and data splitting are all part of a broader set of operations.
Purpose	Verifies that all data is correct, comprehensive, and trustworthy before analysis. Removing noise and irregularities improves the reliability of the data.	The data is prepared for analysis or modelling by ensuring it is compatible with algorithms, which improves the performance of the models and makes them easier to understand.
Techniques Used	Handling missing values (e.g., imputation), correcting errors, deleting duplicates, addressing formatting difficulties.	Data transformation (e.g., PCA) is part of it, as is feature selection, feature scaling (e.g., normalisation, standardisation), categorical variable encoding, and data division (train, test, validation).

Aspect	Data Cleaning	Data Pre-processing
Timing in Workflow	Finished first, before any data is prepared for analysis or modelling.	Depending on the context, this might happen before or after cleaning, or it could be part of a broader pre-processing pipeline.
Impact on Analysis/ Modelling	The accuracy and credibility of the outcomes of the analysis and modelling are directly impacted. Conclusions that are biassed or incorrect could be caused by inaccurate data.	Impacts model performance and efficiency. Well-pre-processed data can enhance model accuracy, reduce overfitting, and improve generalization.

Source: - *(Timespro, 2024).*

3.2 ETL (Extract, Transform, Load) Processes

Data warehousing is a big-scale repository that consolidates information from several sources via the extraction, transformation, and loading (ETL) process. The purpose of data preparation for storage, analytics, and ML is to clean and organise raw data according to a set of business rules (Caserta, 2004).

This is achieved using ETL.

The acronym ETL stands for "extract," "transform," and "load," and it describes a data integration process that unifies, cleanses, and organises data from many sources before loading it into a destination system (data warehouse, data lake, etc.)(IBM, 2023).

Data analytics and ML workflows rely on ETL data pipelines as their basis. Data cleansing and organisation are two of the many business intelligence objectives that ETL may meet via a set of rules. ETL isn't limited to monthly reporting or basic analytics; it can also take on more complex problems, including improving back-end operations or end-user experiences. Common uses for ETL pipelines in organisations include:

- Extract data from legacy systems

- Data cleaning is an important step in improving data quality and achieving uniformity.

- Load data into a target database

ETL refers to the steps used to transform data from one system into another. Take a look at the data, make some changes, and then upload it to another location. This idea is often used in data science, databases, and data warehousing. It's a tried-and-true approach of taking data from one place, cleaning it up, and then feeding it into another place for processing. It is possible to monitor and assess ETL as a lifecycle if we see it as a process. Data warehousing and the process of transferring information from several sources to a central repository is another way of looking at ETL. Data is retrieved from one system, converted to a form that can be analysed, and then put into a data warehouse or some other system as part of this process. A "data warehouse" is a centralised database that collects, organises, and makes data from many sources readily available.

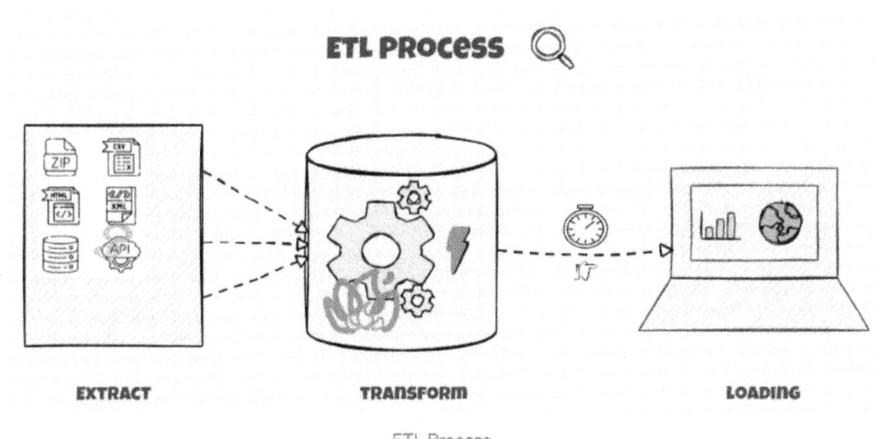

Figure 3.2: ETL process

Sources: - *(Hattatoglu, 2023).*

Fundamentals of ETL Processes

1. Extract — The First Step in Data Mining

Data is "extracted or pulled out" from a variety of heterogeneous sources in the extraction process, as the term implies. A variety of sources, including databases, Excel files, and APIs, may be used. Data that has not been processed yet is prepared for the next step by means of the extraction procedure. Data warehousing and large data transformations often include an extract step, when it is not required to extract all data. In most cases, just the necessary data is retrieved and sent without any mess. Because good data analysis begins with good raw data, it is critical to accurately and thoroughly extract the data. "What you put in is what you get out!" is the data science equivalent of the adage "garbage in, garbage out."

In summary, in this stage:

- Data is retrieved from many source systems, including databases, apps, files, and more.

- There is no room for error or incomplete data collection in the extraction procedure.

2. Transform — The Art of Data

After we have collected or retrieved our data, what comes next? The "Transform" stage is about to begin! In this step, we clean up the raw data so it can be analysed in a way that makes sense. In some cases, we combine data from many sources after cleaning and normalising it. Data transformation to conform to certain criteria, data transformation utilising auxiliary tables and lists, and data combination from various sources are all processes carried out at this level. In data science initiatives, transformation may be an exhausting but ultimately rewarding process. Here you may execute a merging operation, for instance, to ensure that a first name and last name column in a database is mandatory.

In summary, in this stage:

- A data warehousing system-friendly format is applied to the retrieved data.

- Data integration, format changes, rule applications, and data cleaning are all part of the transformation.

3. Load (Loading) — Arrival to Data Warehouse

At last, we reach the loading station! The time has come to analyse our data. How then will we save this information? Here is when data warehouses are useful. During this stage, we transfer all of the cleaned and organised data to a data warehouse so that it can be analysed. Secure and organised data storage made possible by the data warehouse facilitates quick and simple access to the data for analysis. If our data is loadable, then it can be accessed and used for analysis.

To summarise, the data warehouse is the intended recipient of the processed data. Data loading entails transferring information to the data warehouse and arranging it into suitable tables and schemas.

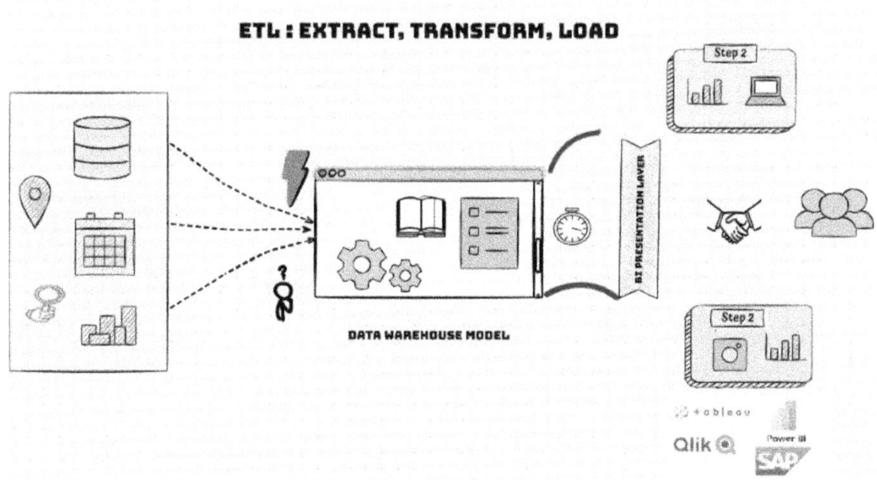

Figure 3.3: ETL

Sources: - *(Hattatoglu, 2023).*

Challenges with ETL

An increasingly difficult aspect of data engineering is creating and maintaining trustworthy data pipelines, even though ETL is crucial due to the exponential growth in data sources and kinds. Constructing pipelines that guarantee data dependability is inherently laborious and time-consuming. Complex programming with low reusability is used to build data pipelines. Data engineers are often in the middle of the bottleneck and have to start from scratch every time since pipelines established in one environment cannot be utilised in another, even if the underlying code is extremely similar. Even more challenging than pipeline creation is data quality management in ever-more-complicated pipeline designs. A data set's value is diminished when bad data is let to run through a pipeline unnoticed. Data engineers must create substantial bespoke code to include quality checks and validation into the pipeline at every stage in order to maintain quality and guarantee trustworthy insights. In conclusion, data dependability becomes very challenging to maintain as pipelines expand in size and complexity, subjecting firms to additional operational pressure in maintaining them. Setting up, scaling, restarting, patching, and updating data processing infrastructure adds time and money to the process. Failures in pipelines are notoriously hard to spot and much more challenging to fix because of inadequate visibility and tools. Any company that aspires to be insights-driven must have dependable ETL regardless of these obstacles. Teams across the company are forced to make choices without trustworthy metrics or reports in the absence of ETL technologies that uphold a standard of data dependability. For data teams to keep growing, data engineers want ETL solutions that simplify the ETL process, empower data engineers to create and use their own data pipelines, and speed up the time it takes to generate insights.

ETL Important

These days, businesses collect data in a variety of formats, including structured and unstructured files, from places like:

- Information gathered from clients' profiles in CRM and online payment platforms

- Inventory and operations data from vendor systems

- Sensor data from Internet of Things (IoT) devices

- Marketing data from social media and customer feedback

- Employee data from internal human resources systems

ETL is a technique that may be used to prepare raw datasets in a way that analytics can make better use of them, leading to more valuable insights. To better manage inventory and anticipate customer demand, online stores might, for instance, analyse data collected at the point of sale. Combining CRM data with social media input allows marketing teams to better understand consumer behaviour.

Benefit of ETL

A more dependable, accurate, thorough, and efficient process is achieved by ETL, which in turn enhances analytics and business intelligence.

1. **Historical context:** Data from the organization's past may be better understood with the help of ETL. Legacy data and data from modern platforms and apps may be combined by a company. You can see the big picture by comparing more current data with earlier records.

2. **Consolidated data view:** Data consolidation is a key benefit of ETL, which allows for more thorough analysis and reporting. Inefficiencies and delays may occur as a consequence of the time and effort required to manage various datasets. ETL integrates disparate data sources and databases into a consolidated image. Data quality is enhanced and time is saved during data integration compared to data movement, categorisation, or standardisation processes. The analysis, visualisation, and comprehension of massive datasets are all facilitated by this.

3. **Accurate data analysis:** To ensure compliance with regulatory requirements, ETL provides more precise data analysis. To make sure the data is reliable, you may combine ETL technologies with data quality tools to clean, audit, and profile the data.

4. **Task automation:** ETL streamlines analysis by automating repetitive data processing activities. You may configure ETL systems to incorporate data changes frequently or even during runtime, and they automate the data conversion process. Data engineers will have more time for innovation and less for routine operations like data preparation and transfer.

ETL Tools

There are four primary types of ETL tools:

- **Batch Processing:** The main ETL procedure used to be batch processing on-premises. In the past, processing big data sets during off-hours was necessary since it taxed an organization's computational capability. Though they are often hosted on the cloud, modern ETL tools have more flexibility in terms of when and how fast they can process batches of data.

- **Cloud-Native.** Data may be extracted and loaded into a cloud data warehouse using cloud-native ETL technologies. After then, they alter the data by using the cloud's power and scalability.

- **Open Source.** An affordable alternative for commercial ETL solutions is open source software like Apache Kafka. While many open-source solutions exist, not all of them are well-suited to deal with data complexity or adapt to new methods of data collecting; for example, some may only help with data extraction. On top of that, open source tools may have a hard time getting support.

- **Real-Time.** Access to data in real-time is essential for modern businesses. To do this, businesses need streaming capabilities and a distributed paradigm for real-time data processing. This is possible with streaming ETL tools, whether they are commercial or open source.

Advantages or Disadvantages:

Advantages of ETL process in data warehousing:

1. **Improved data quality:** The ETL procedure guarantees the accuracy, completeness, and timeliness of the data stored in the data warehouse.

2. **Better data integration:** The ETL process facilitates the integration of data from many systems and sources, increasing its usability and accessibility.

3. **Increased data security:** The ETL procedure may assist to strengthen data security by restricting access to the data warehouse and ensuring that only authorised individuals have access to the data.

4. **Improved scalability:** The ETL process offers a means of managing and analysing vast volumes of data, which may aid in increasing scalability.

5. **Increased automation:** ETL technologies and solutions may streamline and automate the process, saving time and effort when loading and updating data in the warehouse.

Disadvantages of ETL process in data warehousing:

1. **High cost:** ETL process implementation and maintenance may be costly, particularly for resource-constrained organisations.

2. **Complexity:** The ETL process may be complex and challenging to execute, particularly for businesses without the requisite resources or experience.

3. **Limited flexibility:** The ETL method may not be able to handle unstructured data or real-time data streams, which might restrict its flexibility.

4. **Limited scalability**: The scalability of the ETL process may be constrained since it cannot handle vast volumes of data.

5. **Data privacy concerns**: The ETL process might pose data privacy problems since enormous volumes of data are gathered, stored, and analysed.

All things considered, the ETL process is a crucial step in data warehousing that aids in guaranteeing the accuracy, completeness, and timeliness of the data stored there. But it also has its own set of drawbacks and restrictions, so before putting them into practice, businesses must carefully weigh the advantages and disadvantages.

3.3 Real-time Data Processing

Real-time processing, as the name implies, analyses data as it is created and generates results almost instantly. Businesses and organisations across a wide range of industries, from manufacturing and healthcare to banking and e-commerce, depend on this skill.

Key Characteristics of Real-Time Processing:

- **Continuous Data Flow**: Relies on a constant, uninterrupted stream of data flowing into the system.

- **Low Latency:** Minimizes the time delay between data generation and the generation of insights.

- **Immediate Output:** Provides quick responses and actions based on the analysed data.

- **Timely Insights:** Delivers up-to-the-minute insights into current trends, patterns, and anomalies.

Real-Time Processing Differ from Batch Processing

- **Batch Processing:** Collects data over a period of time and processes it in batches at a later point.

- **Real-Time Processing:** Analyses data as it arrives, providing immediate insights and enabling rapid responses.

Feature	Real-Time Processing	Batch Processing
Processing Time	Immediate	Delayed
Latency	Low	High
Data Flow	Continuous	Periodic
Resource Usage	Resource-intensive	Less resource-intensive
Use Cases	Fraud detection, real-time recommendations, autonomous vehicles	Data warehousing, nightly reports, data backups

Real-Time Data Processing Advantages

The following are some benefits that real-time data processing systems provide:

- **Immediate Updating of Databases & Immediate Responses to User Inquiries**: Real-time processing systems guarantee prompt action since there are no processing delays.

- **Up-to-Date Applications:** Real-time processing makes ensuring that fresh data are synchronised quickly in situations where you need to make changes to your apps often.

- **Breakneck Speed:** Fast data loading and storage capabilities are provided by real-time systems, enabling you to process data and derive insights rapidly.

- **Accurate Information:** Your information is always up-to-date. When done in real-time, everything is up-to-date and accurate.

- **Business Agility**: Having current data that reflects the urgency of the market and the changing tastes of consumers makes it simple to respond to a changing business climate quicker and smarter.

- **Quick Detection of Operational Issues:** You may quickly identify bottlenecks and related difficulties with the use of real-time reports that keep you updated on your current company activities.

Real-Time Processing Architecture

The four parts that make up an architecture for real-time data processing are:

- **Real-time Message Ingestion:** Message ingestion systems receive data or messages in streams from many sources and prepare them for consumption by a Stream Processing Consumer. One possible implementation of this service is a folder that just saves new messages. However, a Message Broker is often necessary to act as a message buffer, allowing for dependable message delivery and scale-out processing.

- **Stream Processing:** Stream processors prepare the received messages for analysis or execute data processing activities like filtering and aggregation.

- **Analytical Data Store:** Preparing and managing massive data is Analytical Data Store's forte. Before making data available for examination, they clean it up and organise it so that analytical tools may access it. Analytical data stores are designed to respond quickly to queries and provide sophisticated analytics capabilities.

- **Analysis and Reporting:** Finalising Real-time Data Processing involves making easily accessible and understandable visual representations of the data in the form of reports, graphs, or charts to provide useful, actionable insights.

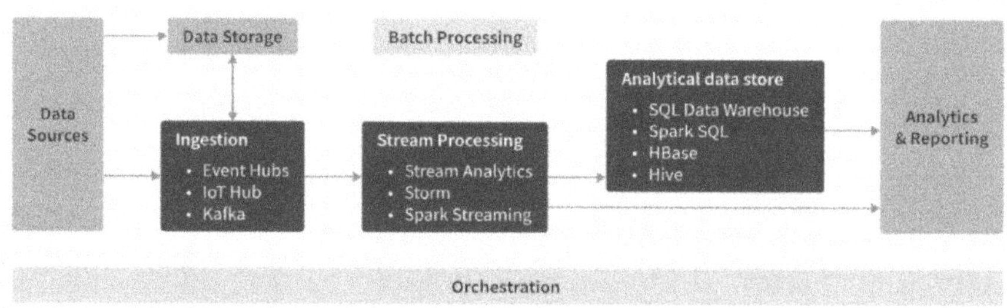

Figure 3.4: Orchestration

Sources: - *(Divyansh Sharma, 2024)*

Tools for Real-Time Processing

- Real-time data processing is already firmly entrenched in several sectors, and it's only getting started.

- Tools that facilitate real-time operationalisation are proliferating in response to the growing demand for such services and the expanding frontiers of real-time analytics and stream processing.

- Businesses may improve their marketing campaigns, sales, and product exposure with the support of real-time data processing solutions.

- Real-time data processing solutions provide several benefits, one of which is the ability to handle small data quantities and produce useful insights rapidly.

- Businesses may benefit from these solutions since they streamline the integration of Real-time Data (RTD) into their systems, allowing for simpler processing and evaluation of data in BI and RTD systems.

- Solutions need to be dependable in their functioning and responsive with accuracy. The list contains a number of possibilities, including

  - Apache Kafka

  - Apache Spark

  - Apache Flink

  - Amazon Kinesis

  - Azure Stream Analytics

  - Google Cloud Dataflow

  - RabbitMQ, and many more.

Challenges to Real-Time Processing

Real-time Processing faces two big challenges:

- **High-Volume Rapid Ingestion**: It is only feasible to handle small amounts of data quickly while processing in real-time. Eliminating bottlenecks in the Data Ingestion Pipeline is crucial to the successful execution of the challenging task of real-time data ingest, processing, and storage. In order to handle data rapidly, the Data Store also has to enable high-volume writes.

- **Speedy Analytics (also called Operational Analytics)**: The ability to generate BI in real-time is the second obstacle to RP system deployment. For businesses to respond swiftly, real-time systems need to provide reports or alerts for staff members as soon as data is available from the source.

Batch Processing

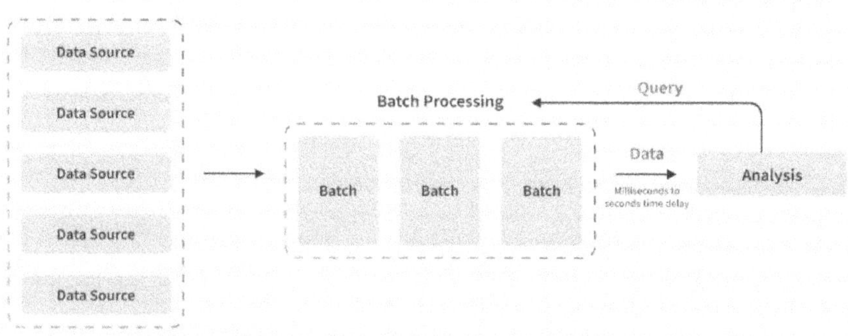

Figure 3.5: Batch Processing

Source: - *(Limeup, 2024)*

- The term "Batch Processing" refers to the act of handling large amounts of data at once. Batch Processing is an alternative to real-time processing that involves gathering transactions at a later date and scheduling their processing.

- Data scientists, analysts, and engineers may use the information shown by the post-processing software application or system to make informed choices based on the data.

- Batch Processing offers more operational flexibility and a delayed response to changing market circumstances compared to Real-time Processing.

- The operation of a washing machine is a straightforward illustration of the application of Batch Processing.

- To use a washing machine, you need to wait for a cloth pile to accumulate before you can wash your garments. This kind of operation is a cost-effective and efficient approach to handle your company's data.

Pros of Batch Processing

- Involve basic systems that don't call for any specialised software or hardware.

- Efficient processing of large volumes of data.

- Works offline.

- Capabilities like automation and minimal user involvement.

- Low maintenance.

Cons of Batch Processing

- Limited in scope and capability.

- No real-time updates.

- Debugging Batch Processing systems is complex.

- May require dedicated staff to handle issues.

Real-Time Processing vs Batch Processing

By consolidating our knowledge of batch processing with that of real-time processing into a single table, we can see the following difference:

Parameter	Real-time Processing	Batch Processing
Time Frame	Predictable reaction times are a hallmark of real-time systems. In milliseconds, they process and consume transactions.	Transactions are recorded and processed at predetermined intervals using batch processing.
Processor Availability	The processor must be responsive continuously because real-time data processing requires zero latency.	Because transactions are scheduled for processing in batches using batch processing, the processor has to be accessible only when needed.

Parameter	Real-time Processing	Batch Processing
Orientation	Processing in real-time is based on events or actions.	Batch Processing is measurement-oriented.
Data Handling Capacity	For smaller data sets, real-time data processing may be very helpful. Massive amounts of computing power and processing capacity are needed for processing massive data sets.	Data quantities that are very huge are best handled via batch processing.
Resources Involved	Computers with high-end architectures and specifications are required for real-time processing.	Standard computer specs should be sufficient for batch processing.
Costs of Implementation	Complex and expensive is real-time stream processing. Quick results are only possible with a certain set of hardware and software.	For corporate applications, batch processing is the most straightforward processing technique. It's also cost-effective.
Maintenance	The complex configuration of real-time stream processing necessitates backup solutions and daily upgrades in order to collect and give data routinely.	Real-time stream processing is more complex than batch processing, which is simpler to handle.

Parameter	Real-time Processing	Batch Processing
Examples	ATM Transactions Data Streaming IoT Sensor or Radar Systems Customer Service Systems	Processing of Payrolls Creation of Billing Cycles Customer Orders Credit Card Transactions

Sources: - *(Divyansh Sharma, 2024)*

Use Cases of Real-Time Data Processing & Analytics

1. **Business Operations**

   - You may optimise your back-office operations with the help of real-time systems that provide your company strategic insights in sales, marketing, customer service, and finance.

   - Faults and operational hazards that might endanger your company's operations can be more easily detected with the use of real-time analytics.

2. **Financial Trading**

   - The capacity to predict when stock prices will increase and decrease is fundamental to trading.

   - The most current information on financial market conditions may be accessed by the Financial Analysts and Traders at your company via Real-time Analytics.

   - They may receive and sync data from social media, databases, news, and weather reports to get a broader view of the market. They will be able to make the best and most educated trading judgements with this perspective.

3. **Customer Support**

   - You can keep your customers' interactions and experiences in sync using real-time systems. Since your Support Team is always up-to-date on your customers' interactions, they can pick up right where the conversation left off.

- A solid foundation for a larger and more loyal client base is laid when customers are treated with the respect they deserve and when their questions are answered promptly.

4.   Marketing Campaigns

- The Marketing Automation Solution provides your teams with real-time data processing, which allows them to see how often customers buy.

- They may utilise the data on consumers' location, age, job title, app use, and a host of other factors to develop more effective ads and useful roadmaps.

- This method assists in developing captivating consumer experiences by accurately targeting your audience.

5.   Sales Initiatives

- Your sales representatives may collect information about potential customers using real-time data processing, operationalise it in your Salesforce/HubSpot customer database, and promptly get in touch with them.

- This improves your sales and conversion rates, which causes your company's income to rise.

6.   Stream Processing

- Stream Processing is a Big Data method that analyses, filters, transforms, or improves data in real time by consuming a continuous data stream.

- Data processing in real-time stream processing systems is time-sensitive because the data moves from origin to destination in real-time.

- Stream Processing Systems consist of many modules that interact with one another and operate in parallel.

Multiple Choice Questions (MCQs)

1. **Which of the following is a common task in data cleaning?**

   a. Removing missing values

   b. Normalizing data

   c. Filtering out outliers

   d. All of the above

2. **What is the primary role of the "Transform" phase in the ETL process?**

   a. Extracting data from different sources

   b. Loading data into a data warehouse

   c. Converting data into a suitable format or structure

   d. Cleaning data for further processing

3. **Which of the following is commonly used for real-time data processing?**

   a. Apache Spark

   b. Apache Flink

   c. Apache Kafka

   d. All of the above

4. **What is the purpose of data validation in the context of data processing?**

   a. To ensure that the data is accurate and correct

   b. To improve the speed of data transformation

   c. To reduce the amount of data processed

   d. To store data in a suitable format

5. **Which of the following techniques is commonly used to handle missing data in datasets?**

   a. Imputation

   b. Duplication

   c. Normalization

   d. Aggregation

6. **In which phase of the ETL process is data typically cleaned, formatted, and aggregated?**

   a. Extract

   b. Transform

   c. Load

   d. None of the above

7. **Which of the following is a challenge in real-time data processing?**

   a. Ensuring data consistency across systems

   b. Reducing data latency

   c. Handling data throughput at scale

   d. All of the above

8. **What is a common method for handling errors during data transformation?**

   a. Skipping the entire batch of data

   b. Logging errors and continuing processing

   c. Deleting all incorrect data

   d. Ignoring data inconsistencies

9. **Which of the following data pre-processing techniques is used to scale numerical values to a fixed range (e.g., 0 to 1)?**

    a. Standardization

    b. Normalization

    c. Aggregation

    d. Binning

10. **Which of the following tools is commonly used to automate ETL processes?**

    a. Apache Hadoop

    b. Apache NiFi

    c. Apache Kafka

    d. TensorFlow

Answer

1	2	3	4	5	6	7	8	9	10
D	C	D	A	A	B	D	B	B	B

Chapter | 04

DATA INTEGRATION AND AGGREGATION

4.1 Data Validation and Error Handling

When building online applications, data validation and error management are crucial, particularly when working with database input and output. They aid in avoiding mistakes, flaws, or hostile assaults that might harm your web app's performance or user experience while also guaranteeing the safety and dependability of your data. Validating and addressing errors in data input and output from your web application's database is something we'll go over in this post.

Data Validation

The term "data validation" describes the steps used to check that data is correct and complete. To apply it, one must include many checks into a report or system to guarantee that the data input and output are logically consistent.

There is little to no human oversight during data entry in automated systems. Therefore, it is critical to check all incoming data for accuracy and quality requirements. There will be major problems with downstream reporting and the data will be useless if input incorrectly. Even with accurate entry, there will be associated expenses for cleaning, processing, and storing unstructured data.

Types of Data Validation

Data validation comes in several forms. When checking data for accuracy before saving it in a database, most validation methods will do at least one of these tests. The following are examples of data validation checks:

1. **Data Type Check:** To make sure the data entered is of the right kind, a data type check is performed. A field may, for instance, only permit numerical entries. The system should reject any data that contains additional characters, such as letters or special symbols, if this is really the case.

2. **Code Check:** The purpose of a code check is to verify that the values entered into a field are legitimate or that the formatting is correct. For instance, comparing a postal code to a database of known valid codes makes the process much simpler. Other elements, such nation codes and NAICS industry codes, may be similarly conceptualised.

3. **Range Check:** Input data may be checked to see whether it falls inside a preset range using a range check. Geographic data often makes use of coordinates such as latitude and longitude. The correct range for a longitude value is -180 to 180 degrees, and the correct range for a latitude value is -90 to 90 degrees. This range is invalid for any values other than this one.

4. **Format Check:** A large number of data types adhere to a standard format. Date columns kept in a defined format, such as "YYYY-MM-DD" or "DD-MM-YYYY," are a popular example of this. Consistency across data and time may be achieved with the use of a data validation technique that checks the format of dates.

5. **Consistency Check:** A consistency check verifies that the data has been input in a consistent and logical manner. You can see this in action when you compare a package's shipping and arrival dates.

6. **Uniqueness Check:** Email addresses and other forms of identification data are inherently distinct. It is expected that these fields in a

database would contain unique entries. To prevent duplicate entries in a database, a uniqueness check is used.

Error Handling

The ability to handle errors is fundamental in data processing and programming. This pertains to the manner in which an application or system reacts when faced with irregularities or surprises while running. Processing errors might involve fixing data that is missing, corrupted, or inconsistent.

Functionality and Features

If an error or problem arises while a program or process is running, Error Handling may catch it and fix it. To keep a system or application functioning smoothly, it uses several tactics including handling exceptions, detecting anomalies, and inspecting logs. Syntax and logical mistakes are both handled via error handling.

- **Exception Handling:** This entails handling exceptions and faults that occur while a system is running in order to prevent the system from failing.

- **Anomaly Detection:** This function finds anomalies, or really uncommon data items, in a dataset that may hint to mistakes.

- **Log Inspection:** Finding, tracking, and fixing mistakes requires looking through system and process logs.

Benefits and Use Cases

A system or application's robustness, resilience, and dependability are enhanced via error handling. It improves system availability and user experience by preventing crashes caused by unforeseen occurrences.

To avoid letting inaccurate or inconsistent data taint analysis or forecasts, proper error management is essential in data processing. This is of the utmost importance in situations like real-time data processing, training machine learning models, and making critical business decisions.

Types or Sources of Error

Logic, run-time, and compile-time errors are the three main categories:

1. When programs behave erratically but don't crash, it's called a **logic error**. Even if it doesn't seem like a logic problem at first, it might be causing unexpected or undesirable results or actions.

2. Invalid input data or unfavourable system settings are the most common causes of **run-time errors**, which occur while a program is being executed. A logical mistake, insufficient memory, or a memory clash with another program are all examples of such issues. When the desired outcome is not achieved when the code is performed, this is called a logic mistake. Extensive program debugging is the gold standard for handling logic problems.

3. Prior to the program's execution, at **compile-time**, mistakes tend to increase. For instance, the program may not be able to compile due to a syntax mistake or a missing file reference.

Classification of Compile-time error –

1. **Lexical:** Identifiers, keywords, and operators that are misspelt fall under this category.

2. **Syntactical:** a missing semicolon or unbalanced parenthesis

3. **Semantical:** operator and operand types do not match or do not agree on the assignment of values.

4. **Logical:** code not reachable, infinite loop.

Advantages of Error Handling in Compiler Design:

1. **Robustness:** Enhancing the compiler's strength, mistake handling enables it to gracefully handle and recover from many types of errors. This ensures that, even when errors are visible, the compiler may continue processing the data program and provide meaningful error messages.

2. **Error location:** A compiler is able to detect and identify errors in source code by combining error handling components. This includes typos, grammatical errors, semantic flaws, type errors, and anything else that might cause the program to behave unexpectedly or provide an incorrect result.

3. **Error revealing:** Clients and software engineers may benefit from compiler error handling services that provide practical error responses. It generates interesting error messages that help developers understand the kind and location of the error, which in turn allows them to address the problems more efficiently. In the troubleshooting system, designers save a lot of time with clear and precise error messages.

4. **Error recuperation:** The compiler is able to recover from errors and continue the aggregation cycle whenever conditions permit thanks to mistake handling. Various techniques, including as error correction, error synchronisation, and resynchronisation, are used to achieve this. To avoid an abrupt termination of the conversation, the compiler attempts to fix the mistakes and goes on with assembly.

5. **Incremental gathering:** The ability to organise and execute correct program segments regardless of whether other segments contain errors is made possible by mistake handling, which enables progressive aggregation. Because it lets engineers test and explore specified modules without recompiling the complete source, this aspect is particularly useful for projects with vast scale.

6. **Productivity improvement:** Legitimate mistake handling allows the compiler to reduce the effort and time needed for error correction and troubleshooting. Helping programmers quickly identify and fix errors, it supports blunder recovery and provides accurate mistake notifications, leading to improved efficiency and shorter development cycles.

7. **Language turn of events:** An essential aspect of language planning and development is error correction. Consolidating error handling

in the compiler allows language designers to define typical error behaviour and authorise clear requirements and standards. This ensures that developers adhere to the anticipated use designs, which in turn increases the language's overall reliability and consistency.

Disadvantages of error handling in compiler design:

1. **Increased complexity:** Error handling in compiler design can significantly increase the complexity of the compiler. This can make the compiler more challenging to develop, test, and maintain. The more complex the error handling mechanism is, the more difficult it becomes to ensure that it is working correctly and to find and fix errors.

2. **Reduced performance:** Error handling in compiler design can also impact the performance of the compiler. This is especially true if the error handling mechanism is time-consuming and computationally intensive. As a result, the compiler may take longer to compile programs and may require more resources to operate.

3. **Increased development time:** Developing an effective error handling mechanism can be a time-consuming process. This is because it requires significant testing and debugging to ensure that it works as intended. This can slow down the development process and result in longer development times.

4. **Difficulty in error detection:** While error handling is designed to identify and handle errors in the source code, it can also make it more difficult to detect errors. This is because the error handling mechanism may mask some errors, making it harder to identify them. Additionally, if the error handling mechanism is not working correctly, it may fail to detect errors altogether.

4.2 Overview

An essential part of data management is integrating and aggregating data, two separate but complementary steps in getting data ready for analysis and decisions. Data integration is the process of bringing together disparate datasets in order to get a more complete picture. Data cleansing, transformation, and unification are all parts of this process, which aims to make different information accessible and harmonised. Creating a unified data environment that can allow in-depth analyses is the main objective. In contrast, data aggregation seeks to provide broad insights by gathering and synthesising data. The objective is to merge data from several sources into a single, usable dataset for analysis and reporting. This approach simplifies complex data so that it may be used for strategic decision-making.

4.3 Data Integration Techniques

A modern data strategy or process must have data integration as one of its core components. Structured databases and unstructured text files are two examples of the many forms that data may take. Making sense of all this data might seem like an insurmountable mountain to climb. Luckily, this process can be made more effective and streamlined with the use of innovative data integration solutions.

Types of Data Integration Techniques and Strategies

Data integration methods and techniques abound, so you can be certain that your data will be successfully extracted, transformed, and loaded. Several methods exist, therefore let's examine them:

1. Data Consolidation

The process of data consolidation is merging information from many sources into one larger database, warehouse, or lake. Because it creates a unified view of data for reporting and analysis, this approach is great for companies dealing with a varied data environment.

Data kept in several databases across the company is one example. You may save time and effort compared to manually querying each source by storing all of this data in one central repository.

2. Data Federation

Virtual data integration, or data federation, lets businesses access and analyse data in real-time from several sources without actually transferring or duplicating it. Data federation is an alternative to centralising data storage that allows users to access data from several sources via a single virtual layer.

Consider the following scenario: you may have three databases housing different types of data: one for customers, one for sales, and one for inventory. The goal of data federation is to provide a uniform picture of all your data by making it accessible via a single interface. Examples of common data federation use cases include analytics for customers, suggestions for products, and detection of fraud.

3. Data Transformation

Data transformation is the process of converting raw data from one format to another, with the goals of making it more usable in subsequent operations by cleaning, normalising, and otherwise improving the data. To make them usable for analysis or modelling, dates may need to be reformatted or text strings converted to numerical values.

An example would be a CSV file including client information. The dates will need to be reformatted and the text strings converted to numbers before they can be loaded into your database.

Transferring data from one system or application to another also requires data transformation. To properly load data from one database into another, for example, you must first verify that the data is in the proper format.

4. Data Propagation

The process of copying data from one location to another is called data propagation, data replication, or ETL (extract, transform, load). Both real-time (synchronous) and planned intervals (asynchronous) execution are possible. When data has to be shared across several systems and places, this approach comes in handy.

A cloud-based software, for instance, could need access to client information kept in an on-premises database. Through data propagation, you may transfer data from an on-premises system to a cloud storage, so your app can access it without directly querying the source.

The process of data propagation is also utilised when transferring and converting data from several sources into one main database. Having a copy of the data in the central location makes analysis and access much simpler. Big data, consumer analytics, and reporting are some of the most common applications.

5. Middleware Data Integration

As go-betweens, middleware solutions make it easier for different systems to exchange and process data. Their ability to provide smooth data interchange and translation makes them indispensable in intricate integration situations.

You may, for instance, have two systems that store client information differently; one may use XML while the other would prefer JSON. Avoiding the need to manually code the transformation process, middleware makes it easy to convert between multiple formats.

Furthermore, middleware solutions may provide capabilities like automatic processing, real-time synchronisation, and assured data transmission among the two systems. Middleware solutions enable organisations to power complex integration scenarios, including system-to-system interface and enterprise application integration (EAI).

6. Data Warehousing

Data warehousing is the practice of systematically storing and organising data in a single location. Particularly helpful for reporting and business intelligence, it gives data a historical context.

For instance, it's not uncommon for many databases throughout the company to hold sales data. By storing everything in one place, you'll have easy access to your sales history whenever you need it, whether for reporting or analysis. Data warehousing also makes it feasible to do complicated analytics that would be very difficult, if not impossible, to accomplish with conventional database queries.

When it comes to storing and analysing massive volumes of data, data warehouses are a must for every organisation. Financial reporting, predictive modelling, and consumer analytics are some of the most common use cases.

7. Manual Data Integration

Data extraction, transformation, and loading from several sources may sometimes need human intervention in manual data integration. Although this approach requires a lot of resources, it works well for smaller integration projects.

For instance, you could have to glean client information from a wide range of places, including online forms, surveys, and social media postings. If the data is not already accessible in a structured manner, manually integrating it can be the only other alternative, although it can be rather time-consuming.

Companies resort to manual integration when dealing with complicated or sensitive data, which necessitates human interaction. Customers can be categorised, fraud can be identified, and compliance may be achieved.

Technique	Description	Use cases
Data consolidation	Organises information from several sources into one central registry.	Customer 360-degree vision, data warehousing, and master data management.

Technique	Description	Use cases
Data federation	Makes it possible for users to obtain data from several sources without transferring or copying any data.	Analytics for customers, suggestions for products, and detection of fraud.
Data transformation	Transforms data between different formats.	Data preparation, data transfer, and data analysis.
Data propagation	Copies information across locations.	Backup of data, real-time analytics, and data warehousing.
Middleware data integration	Performs data integration via middleware solutions.	Interaction between systems, enterprise application integration (EAI).
Data warehousing	Maintains data storage in a single location.	Big data, financial reporting, and forecasting from customers.
Manual data integration	Integrates data by hand.	Minor undertakings, undertakings with unstructured data.

Source: - *(Chen Cuello, 2024)*

Popular Data Integration Technologies

Let's examine the most often used tools and technologies for data integration now that you are aware of the various approaches. These include:

- **Extract Transform Load (ETL):** Transferring and transforming data across systems is the primary function of ETL tools. With their help, users may gather data from many sources, convert it to their preferred format, and then import it into the intended system. A few examples of well-known ETL systems include Oracle Data Integrator, River Data Integration, Talend Data Integration, and Informatica PowerCenter.

- **Enterprise Information Integration (EII):** EII technologies bring together data from several sources to create a cohesive picture. Users may access and query data without physically moving or copying it thanks to this. Oracle Database Integration Services, TIBCO Active Matrix Business Works, and IBM WebSphere Information Integrator are some of the most well-known EII systems.

- **Application Programming Interfaces (APIs):** Applications and systems may be linked by APIs, eliminating the need for human coding. By creating a barrier between disparate systems, they facilitate the creation of seamless connections with little effort on the part of developers. Many businesses use API solutions from popular providers like Google Cloud, Amazon Web Services, and Microsoft Azure.

- **Enterprise Data Replication (EDR):** It is possible to duplicate data across several sites with the help of EDR solutions. They let users keep their data consistent and current across several platforms or databases. Oracle Golden Gate, Informatica Power Exchange, and IBM Infosphere Change Data Capture are some of the most well-known EDR systems.

Data Visualisation

To display data graphically, data visualisation technologies are used. This facilitates the user's ability to swiftly comprehend large data sets and spot important trends or patterns. Tableau, Microsoft Power BI, and QlikView are three well-known data visualisation systems.

Which data integration strategy is right for your business?

The demands and requirements of your company will dictate the data integration approach that is most suited to your needs. Everything from the data's complexity to the integration points you're making to the analytics you want to do should be carefully considered.

A middleware solution might be the best option if you need to combine massive volumes of data from several sources. However, an ETL solution can be the way to go if all you need to do is transfer structured data from one system to another.

Data velocity and volume, latency needs, budget, and system complexity are all important considerations when deciding on the best approach for your company.

Businesses with varied data integration requirements may potentially benefit from a hybrid strategy that combines many approaches.

If you want to make a smart choice, it's a good idea to consult with data integration specialists and evaluate your needs thoroughly.

Integration Strategy	Suitable for	How it Works	Pros	Cons
Batch ETL	Non-real-time integration	Extract, Transform, Load in batches	Cost-effective, suitable for large data	Non-real-time, potential data latency
Real-time ETL	Near real-time integration	Continuous extraction, transformation, loading	up-to-the-minute information, essential for making decisions rapidly	More complex and potentially expensive
Change Data Capture (CDC)	Real-time data synchronization	Records and reproduces changes made to the source system	Real-time updates, minimizes data latency	Complex setup, resource-intensive
Data Federation/ Virtualization	Heterogeneous data sources	Offers a consolidated perspective without the need for data acquisition	Minimizes data duplication, simplifies access	Performance challenges for complex queries

Integration Strategy	Suitable for	How it Works	Pros	Cons
Data Replication	Distributed data synchronization	Copies data between systems, maintains synchronization	Ensures data consistency across locations	Resource-intensive, potential data conflicts
API-Based Integration	Third-party services/cloud apps	Connects systems via APIs for data exchange	Efficient for cloud services and external partners	Possible need for bespoke development due to lack of command over third-party APIs

Source: - *(Chen Cuello, 2024)*

The future of data integration

New approaches that take into account the present economic situation and solve the problem of a lack of qualified data engineers will shape the data integration landscape going forward. Here are some important trends and concepts that will shape the future of data integration:

- **Automation and Efficiency:** Platforms that automate ETL procedures are becoming more popular among businesses as a means to reduce the need for human data preparation and to save resources.

- **Flexibility and Adaptability:** Systems that are able to adapt and provide ETL flexibility are essential for dealing with varied data integration difficulties, since data sources and needs are always changing.

- **Unified Data Views:** Making educated judgements about operations, customers, and markets is made possible when organisations integrate data from many sources into one holistic picture.

- **Control and Transparency:** Improving the ability to manage data flows and operations is a key component of future data integration solutions. This will enable better cross-departmental communication and the assurance that choices are grounded in reliable information.

- **Real-time Data Movement:** The ability to integrate data in real-time is crucial for making quick strategic choices in today's fast-paced environment.

- **Time-to-Insights Optimization:** Businesses may respond quickly to changes in the market by speeding up ETL pipeline transformations, which shorten the time it takes to get insights from data.

- **Data Activation for Decision-Making:** The main goal is to facilitate organisations' data utilisation by making data easily accessible for analysis and decision-making.

- **Actionable Insights:** The focus is on reducing the time it takes to get from collecting data to making a decision by converting it into actionable insights as soon as possible.

- **Cost Savings and Value:** In order to save money, increase productivity, and minimise data-related mistakes, businesses are looking for affordable data integration solutions.

- **Adapting to Change:** For organisations to maintain a competitive edge, platforms must be able to effortlessly adjust to evolving data sources, technologies, and business requirements.

- **The rise of cloud-based data integration solutions:** The many benefits that cloud-based data integration solutions provide over their on-premises counterparts are driving their rising popularity. Among these benefits include the capacity to scale, adaptability, and low cost.

- **The increasing use of artificial intelligence and machine learning in data integration:** Automating data integration processes, improving data quality, and discovering new insights from data are all possible due to AI and ML.

- **The growing importance of data security and privacy:** The significance of data privacy and security is rising in relation to the amount of data that organisations gather and keep. Safeguarding data from misuse or unauthorised access is an essential feature of data integration systems.

4.4 Data Aggregation Strategies

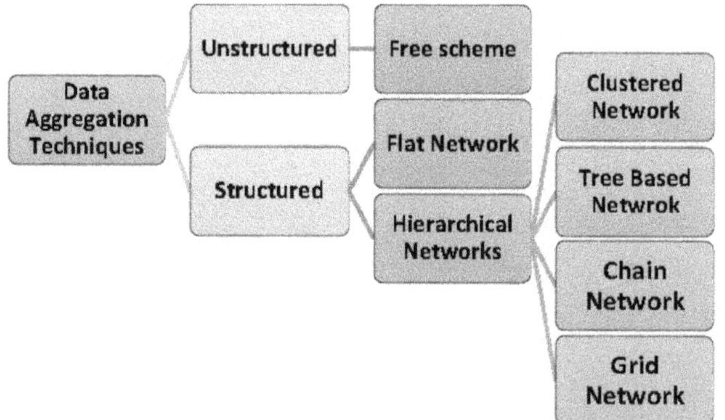

Figure 4.1: Data Aggregation Techniques

Source: - *(Abdulzahra & Al-Qurabat, 2023)*

Aggregation of data refers to gathering information and displaying it in a condensed form. Both the quantity and quality of the data acquired determine how accurately the results can be analyzed and communicated. Thus, a vital part of any literature evaluation is data aggregation.

The statistical analysis and summarisation of collected research data are two outcomes of data aggregation. The business sector has long acknowledged the value of data summarisation as a method of information gathering, organisation, and presentation.

How is Data Aggregation Done?

Common data aggregation technologies include Google Looker, Zoho Analytics, Cloudera Distribution, and many more. The term for this is data aggregators. Typical characteristics for these include the ability to gather, analyse, and display aggregate data.

Your credit score provider Experian is just one example of a data aggregator that has collected hundreds of pieces of information on individual customers. Now, that data can be exploited for highly targeted marketing initiatives, going beyond just providing a basic three-digit score.

Data Aggregation Techniques

There is an efficient method that may be followed to aggregate data utilising four different strategies.

1. **In-network Aggregation:** In a multi-hop network, this is the standard procedure for collecting and directing data.

2. **Tree-based Approach:** Building an aggregate tree is the starting point for the tree-based method's definition of aggregation. With a minimum spanning tree structure, the sink node is like a root and the source node is like a leaf. Data development begins at the leaf node and progresses to the root node, which is also known as the base station.

3. **Cluster-based Approach:** To compile more extensive data over the whole network, a cluster-based technique is used. It is categorised into a few of groups. Members of each cluster vote for one person to serve as cluster leader.

4. **Multi-path Approach:** In the multi-path technique, data is transmitted to a single parent node, often known as the "root node" in the aggregation tree. From there, data may be transferred via numerous pathways. In this network architecture, every node has the capability to deliver data packets to its many inputs.

Data Mining

Data aggregation only meets half of the demands of potential users. One big database with so much information might potentially save a significant amount of time compared to working with many separate databases. That time can only be saved, however, if the compiled data can be mined or searched to swiftly and precisely locate the information you need.

- **Garbage in, Garbage out (GIGO)**

  These huge data warehouses seem to be living proof of the old adage that the quality of the data you get from a computer is directly proportional to the quality of the data you input into it. If your library does not have access to restricted-access databases like Academic Search Premier, you will have to spend a lot of money to get the most up-to-date research. Journal Storage, or JSTOR for short, provides limited free access to current journal releases along with alternative subscription choices.

- **Open Access Databases**

  A response to the perceived elitism of subscription-only research publications, Open Access Publishing is devoted to making research data freely available. Databases like High Wire at Stanford University and the Public Library of Science (PLoS) have expanded substantially because of this initiative. For instance, High wire ranks among the top open access databases with about eight million articles and over 2.5 million free full-text articles; nevertheless, many other databases fall short of these benchmarks:

- While there are hundreds of thousands of articles available on the Social Science Research Network (SSRN), a large number of them are "working papers" that were written before they were published and may need further communication with the authors.

- There are more than 24 million papers available in the NIH's PubMed database, but access to many of them is limited.

- Google Scholar uses the 'Mother Google' search algorithm, but it returns a lot of restricted articles and journals that haven't been peer reviewed, so their reliability is in doubt.

4.5 Ensuring Data Consistency

The term "data consistency" describes how dependable, accurate, and consistent the information is that is saved and processed in a database. It keeps data intact and valid throughout time by making ensuring it doesn't alter during operations and transactions. Data contained in databases should also be consistent so that they appropriately represent happenings in the actual world. By making sure that all data copies are same across all databases and systems, it avoids data inconsistencies and conflicts. The significance of data's consistency, precision, and integrity cannot be emphasised enough in an era when data is valued more than oil.

The importance of data consistency

Consistency in data is important for several reasons. Users are able to obtain accurate and up-to-date information whenever they need it, which improves their ability to make business choices. Also, businesses may improve productivity, cut down on mistakes, and simplify processes with consistent data.

- **Trustworthiness:** Users, consumers, and stakeholders have confidence in a system when data is consistent. User confidence in the system's information correctness is enhanced when data is consistently dependable and unverified. Connections with clients and business associates thrive when trust is there. Collaborations, partnerships, and transactions are more likely to occur when stakeholders have faith in the information a company offers.

- **Decision Making:** The ability to make educated judgements at every level of an organisation relies on data that is accurate and consistent. choices at all levels, from strategic planning to operational choices and tactical actions, benefit from precise and trustworthy data. Better analysis and less likelihood of poor judgements leading to lost opportunities, inefficiencies in operations, or financial losses may be achieved via the use of more accurate and unified data.

- **Compliance:** Data consistency and integrity regulations are very strict in several fields, including healthcare, banking, and the public sector. It is essential to comply with these standards in order to preserve confidence and credibility, as well as to comply with the law. Fines, legal penalties, and harm to one's reputation are among the serious outcomes that may result from not following regulatory requirements.

- **Customer Satisfaction:** Customer experiences are enhanced when accurate and trustworthy data is provided across all touchpoints. Organisational trust and customer satisfaction are enhanced when customers obtain consistent information about goods, services, pricing, and availability. Customer trust and happiness may be negatively impacted by inconsistencies like price disparities or inconsistent product descriptions, which can result in bad impressions and even revenue loss.

What Causes Data Consistency Issues?

There are a number of potential causes of data consistency difficulties, and any of them might compromise the security and trustworthiness of the data stored.

- **Concurrent Transactions:** Concurrent data access is a typical feature of contemporary database systems, particularly in remote or multi-user settings. When several processes or users try to access and change the same data at the same time, it may cause inconsistencies and conflicts. Data anomalies like missing updates, dirty reads, or phantom reads may occur when concurrent transactions conflict with each other and there aren't enough concurrency control methods in place, such locking or timestamp-based protocols.

- **Software Bugs:** Database data may be accidentally changed or corrupted due to programming errors, logic problems, or improper data handling in the computer code. Data truncation, calculation problems, and inappropriate error handling are some ways that software defects might cause recorded information to be inconsistent.

Outdated or inaccurate data sources

Discrepancies may emerge if data sources are either incorrect or out of date. Data sets become unintelligible when they are either out of date or include information from several sources that has not been appropriately harmonised.

- **Lack of integration:** Discrepancies might also arise from data not being adequately integrated across systems. Uneven data distribution across different systems or databases is a common problem. This may happen as a result of data silos or because integration technologies were applied incorrectly.

- **Hardware Failures:** Database operations and data consistency are both jeopardised when hardware fails or malfunctions, such as when discs crash, memory issues occur, or networks go down.

Data corruption or loss may occur during write operations due to sudden power outages or hardware malfunctions. To restore consistency, recovery measures like journaling, mirroring, or replication are necessary.

- **Human Error:** Inconsistencies in the database system may be introduced by users, administrators, or developers making mistakes. Incorrect data entry, improper operation execution, and improper database configuration are common examples of human error. Human mistake, including insufficient training, carelessness, or exhaustion, might undermine data consistency.

- **Lack of Validation:** Inadequate data validation and integrity checks pose a threat to database consistency and correctness by allowing inaccurate or incomplete data to be kept. Data submitted into the system could include mistakes, inconsistencies, or unauthorised values if the appropriate validation procedures are not in place, which might cause difficulties with data quality and downstream processes.

Steps to Help Ensure Data Consistency

The following steps may be taken by organisations to ensure data consistency:

- **Transaction Management:** To guarantee consistent and reliable execution of database transactions, use transaction management techniques such as ACID (Atomicity, Consistency, Isolation, Durability) attributes. Make sure that no two transactions may access the same data at the same time by using locking techniques.

- **Data Validation and Constraints:** Ensure that only accurate and consistent data is kept by enforcing data validation rules and constraints at the database level. In order to keep connections between data entities intact, you must establish referential integrity rules and foreign essential constraints.

- **Regular Data Consistency Checks:** Find and fix database errors and discrepancies by conducting audits and data consistency tests. To ensure data integrity and proactively detect problems, use scripts and automated tools.

- **Backup and Recovery Procedures:** To safeguard data from corruption or loss due to hardware failure, natural catastrophes, or other unanticipated occurrences, establish solid backup and recovery protocols. Make sure your backup and recovery procedures are effective and reliable by testing them regularly.

- **Data Governance and Documentation:** The organization's data generation, consumption, and maintenance should be governed by well-defined rules and processes. The consistency and integrity of data, as well as its comprehension and compliance, may be guaranteed by documenting data definitions, standards, and procedures. Important business operations and decision-making activities rely on consistent data quality.

Multiple Choice Questions (MCQs)

1. **Which of the following is a common data integration technique?**

    a. Data replication

    b. Data transformation

    c. Data federation

    d. All of the above

2. **What is the primary goal of data aggregation in data engineering?**

    a. To combine and summarize large sets of data into a useful format

    b. To extract data from different sources

    c. To transform raw data into structured data

    d. To clean data by removing duplicates

3. **What is the main challenge in ensuring data consistency in distributed systems?**

    a. Dealing with network failures

    b. Ensuring all systems have the same version of data at all times

    c. Reducing data volume

    d. Handling data in unstructured formats

4. **Which of the following techniques is used to combine data from different sources while maintaining access to the original data in real time?**

    a. Data federation

    b. Data warehouse integration

    c. Data replication

    d. Data cleaning

5. **In data aggregation, what is the purpose of performing a "group by" operation?**

    a. To filter irrelevant data

    b. To summarize data based on specific criteria

    c. To sort the data

    d. To join data from multiple sources

6. **Which of the following techniques is commonly used to maintain data consistency in a distributed database system?**

    a. Horizontal scaling

    b. Consistent hashing

    c. Two-phase commit

    d. Sharding

7. **Which of the following is a key benefit of using an Extract, Load, Transform (ELT) approach over ETL in data integration?**

    a. Data is transformed before loading, improving the speed of processing

    b. Raw data is loaded into the data warehouse first, allowing for more flexibility in transformation later

    c. ELT does not require data validation

    d. ELT is not used in data integration

8. **Which of the following is a common type of data aggregation used in time series data?**

    a. Average aggregation

    b. Count aggregation

    c. Sum aggregation

    d. All of the above

9. **Which database consistency model ensures that all transactions are processed in a manner that guarantees data consistency, even in the event of a failure?**

    a. Eventual consistency

    b. Strong consistency

    c. Read consistency

    d. Tuned consistency

10. Which of the following tools is used to facilitate data integration and enable real-time data movement?

a. Apache NiFi

b. Apache Hadoop

c. MongoDB

d. PostgreSQL

Answer

1	2	3	4	5	6	7	8	9	10
D	A	B	A	B	C	B	D	B	A

Chapter 05

DATA QUALITY AND VISUALIZATION

5.1 Defining Data Quality Metrics

Metrics for data quality are indications that teams use to gauge the quality of data. With their assistance, we can distinguish good data from bad data. These are just a few of the numerous advantages, however. There are other benefits to measuring data quality:

- **Identifying areas for improvement** – In order to improve product quality and performance, quality measurements reveal problems like a high failure rate or a low customer contentment rating.

- **Understanding process effectiveness** – Monitoring task or delivery timelines are examples of quality metrics that teams may use to assess the efficacy of their present procedures.

- **Developing solutions** – Assessing quality allows you to fine-tune your strategy in sectors like manufacturing. For instance, you might invest in various materials to minimise mistakes.

- **Cost reduction** – Rework, more QA, and the expense of client warranties are all ways in which data quality concerns drive up costs. You can save costs in the long run, improve product quality, and identify issues earlier by analysing quality measurements.

Data quality dimensions vs. data quality metrics vs KPIs

Relevant and sometimes sharing common causes are data quality dimensions, which group related data quality challenges into manageable groups. Metrics for data quality describe the explicit quantitative or qualitative measurement of a dimension and how it may be tracked over time. Key performance indicators (KPIs) with high data quality show how well you're doing in reaching your business goals.

Metrics to assess data quality

1. Data to errors ratio

A straightforward method for evaluating data quality is offered by this ratio. It entails counting all the instances of data mistakes in a dataset, including missing, incomplete, or duplicate items. After that, it is compared to the overall size of the dataset.

Data quality improves when fewer errors are found while the data amount stays the same or grows.

This strategy's drawback is the possibility of unintentional data mistakes. When this is included, it becomes problematic to utilise the data-to-error ratio alone, since it may provide an overly optimistic view of the quality of the data.

2. Number of empty values

Empty values often indicate that important information was entered in the wrong field or is missing. This is a very easy data quality problem to monitor. All you have to do is find out how many items in a data collection contain blank fields, and then monitor that amount over time.

It seems to reason to prioritise data fields that significantly increase the overall value. An optional remark field, for instance, might not serve as a reliable indicator of data quality. However, the overall completeness of data sets is enhanced more by an essential item, such a phone number or zip code.

3. Data transformation errors

Data quality problems are often indicated by problems with data transformation, which is the act of converting data from one format to another. The transformation will most likely fail if a required field has a null or unexpected value that doesn't match business needs. The frequency of failed or incomplete data transformation operations is a good indicator of the data's overall quality.

4. Amount of dark data

"Dark data" refers to information that your business acquires and keeps but never uses. No one ever looks at dark data, thus it's a good sign that there are data quality concerns. To this day, many businesses fail to see the full value of the data they already have. Now is the time to find the data, get it out of storage, and make sure it is accurate, consistent, and comprehensive.

5. Data storage costs

Does the quantity of data you consume stay the same, but the price of data storage keeps going up? A lack of high-quality data is usually the cause of this.

Problems with data quality could arise if you save information without putting it to use. However, you can be sure that you are improving data quality if you see a decline in storage costs with steady or growing data activity.

6. Data time-to-value

How fast is your team able to turn data into value for the company? You could learn a lot about your data's quality in general from the answer.

There may be an issue with the data quality if data transformations make a lot of errors or if human intervention and cleaning are needed. As a result, you should immediately begin working on a robust data quality methodology.

7. Email bounce rates

An effective email list is the backbone of every successful marketing or sales campaign. Data from customers and prospects could deteriorate rapidly, leading to inaccurate datasets and ineffective marketing efforts.

Inadequate data quality is a common cause of email bounces. They happen when data is inaccurate, missing, or out of date, and emails are sent to the wrong addresses.

8. Cost of quality

At last, there's a single statistic that all teams who put effort into data quality management may use.

The golden grail for every data practitioner is having access to high-quality data. Although data set polishing isn't our primary concern, we all work for companies whose primary goal is to increase revenue and expand their company.

Being able to demonstrate the return on your efforts in high-quality data is crucial. Is it possible that certain measures for measuring data quality are more important than others? Your organization's particular needs will always determine which KPIs you prioritise.

How to improve data quality using data quality metrics?

Although there are many various types of poor-quality data, there are a few common characteristics that you may concentrate on when trying to improve data quality. People, Process, Technology (PPT) is a helpful framework for organisational operational improvement.

These best practices, which come from this framework and others, can help you use the data quality measurements you have been monitoring to enhance the quality of your data.

Hold your team accountable for data quality metrics

How can you and your team members work together to establish data quality measures, ensure that the organisation meets those standards, and provide data of the highest quality?

To ensure the integrity of your downstream data products, for instance, the engineering team may use pull request reviews to restrict potentially unstable modifications to upstream systems.

Collaborating with stakeholders outside of the data team—maybe even the business teams—is a good way to assume responsibility for the implementation and enhancement of data quality metrics.

Implement business processes for data quality improvement

Business practices for improving data quality are another key aspect. A data quality evaluation may be run once. Alternatively, you may include metric scorecards and data quality measurements into quarterly OKRs to concentrate on continuous development.

Additionally, data entry training sessions may greatly help the sales and marketing staff, which in turn can affect the authenticity and integrity of the data.

Use data quality tools

Lastly, it's recommended to use technological solutions that assist in enhancing data quality throughout its whole lifespan. Solutions for gauging data quality and data cleaning tools for adjusting data values abound in the data quality tools industry.

How to put data quality metrics into practice

1. Identify your unique pain points around data quality

Which issues with data quality have recently presented the most obstacles for your business? Maybe you can identify anything particular, such as:

- The creation of a sales forecast does not update customer data.

- Business users find it challenging to understand dashboards.

- The most current business indicators are not being used in the calculations.

Gaining a clear picture of your problems allows you to choose the best strategies for fixing them and improving your data quality.

2. Define use cases specific to your organization

In the last point, you identified your pain spots; this one is linked to that. Having a clear understanding of your objectives for certain data use cases is just as important as describing your data difficulties.

Here are a few examples:

- Providing updated reports based on comprehensive data is your objective in this use case, which is sales forecasting for quarterly board meetings.

- Building an accurate and comprehensive database of client information is your objective in this use case, which is to bolster marketing campaigns centred around emailing.

This is the best way to inform the right people about the importance of data quality measurements and how to link them to what really matters.

3. Connect to data quality dimensions

The next step is to link the data quality characteristics you've chosen to the measurements you'll use to evaluate them. The aspect of punctuality is a good case in point. How will you determine its value? Consider the following metrics as examples:

- The average duration between the time the dashboard was accessed and the time the data was last refreshed within it

- The executives are asked to judge the usefulness of the data on their dashboards in a quarterly poll.

- In comparison to a series of tests that autonomously apply business rules such as "the net revenue is more than 0," the dashboard displays a lower number of outputs.

4. Describe how to measure the metrics

Companies may use multiple criteria to evaluate quality depending on their objectives. Teams often follow these guidelines to keep tabs on key metrics:

- **Determine the quality factor to be measured** – So, for instance, you may monitor how long it takes to reply to customer enquiries or how many error messages a customer gets. Given that the findings could suggest adjustments to criteria, procedures, or goods, it's important to think about controllable aspects.

- **Determine the scope** – Metrics for data quality may be monitored for a specific feature or for a whole suite of projects. Maintaining the data quality initiative's focus requires early determination of the measurement's scope.

- **Keep track of changes** – Most of the time, the data that is gathered represent a moment in time. You may find out how well your quality measures are working by monitoring changes over time.

5. Make metrics actionable and easy to use

It's great to have metrics for data quality, but it's much better if stakeholders can actually utilise them to have a better understanding of data quality and then take appropriate action. Data quality indicators could be better communicated and presented if they are visually represented. Visualising your data makes your data quality measurements more engaging, easy to grasp, and actionable. In terms of data quality, it may also help you spot trends, outliers, patterns, and irregularities. Figure out which methods of visualising data will help people understand your data quality indicators the best. Bar charts, pie charts, and histograms are a few examples of visual representations that may show how data quality indicators are distributed.

5.2 Data Validation Methods

The goal of data validation is to make sure all the data is correct and consistent. As a result, businesses are more likely to make good use of the massive amounts of data they collect and generate.

Data validation techniques

Data validation involves both engineers and analysts, however their responsibilities may vary. Building and maintaining the data pipeline and developing the tools for data transmission and analysis processes, for instance, may be more likely to be done by data engineers. In contrast, analysts concentrate on translating business challenges into data queries and on the analysis, interpretation, and insights gleaned from data. Their work is mostly focused on the data validation methods discussed below, while their responsibilities may overlap in most organisations.

Data validation techniques for engineers

Some typical data validation methods that engineers could use on a daily basis are listed below.

- **Data type validation:** The information that a data field contains is known as its data type. When validating data types, it's common practice to look for numbers in numerical data fields and discard any other characters, including capital letters or special symbols.

- **Range validation:** For example, the numbers 1 through 10 make up a range, which also includes the extremes of the range as well as the values in the middle. When using range validation, you can only accept values that fall within the specified range. Validating the range from -20 to 40 degrees might help you catch a value of -25 degrees in your data.

- **Format validation:** Part of format validation is making sure there's a predetermined format. This is shown, for instance, by comparing the date formats MM-DD-YYYY and YYYY-MM-DD. When working with data from several nations, this might be important to keep in mind since date format conventions change. Data that doesn't conform to the format is identified and corrected via validation.

- **Presence check:** A presence check just verifies that a field does not contain any empty values. Data sets that are lacking names may not provide essential information. This might become problematic, however, if the last name is lacking. Checking for presence guarantees that all viable datasets have data in the last name column.

- **Pattern matching:** In situations involving massive amounts of data that may include human data input, pattern matching proves to be quite useful. To ensure that all data in the collection follows a certain pattern, it verifies against that pattern. A standard email format comprises the following elements: a domain name followed by a.com,.org, or other top-level domain, and some content before the @ symbol. Data that doesn't conform to this pattern will be marked as invalid, for example, an email address that is lacking a @ symbol.

- **Cross-field validation:** The process of cross-field validation involves comparing and checking two or more fields in relation to one another. Consider a comparison between the overall number of tickets sold and the sum of the number of concertgoers with VIP, premium, and general admission tickets. If there are issues with the ticket count or a mistake in the total number of seats available in a stadium, engineers will be notified of this.

Data validation techniques for analysts

- **Uniqueness check**

  A uniqueness check is exactly what it says it is. For example, in a list of workers' Social Security numbers or a list of shop orders to be sent out, this validation checks to determine whether each data item is really unique. Something is amiss if the data in any two fields is identical.

- **Data profiling**

  Data profiling aims to comprehend three data aspects simultaneously and is more complicated than other data validation methods:

  - Structure, or data organization and format

  - Content errors and inconsistencies

  - The connection between several pieces of information, such the information in two different tables in a dataset

  The proper format of data is essential for data analysis and gaining business insights; data profiling helps in this endeavour.

- **Statistical validation**

  Data is assessed in statistical validation based on inferences drawn from it. The ability to reproduce a scientific research using just its data is one of its primary benefits. Statistical validation ensures that the data results are legitimate and in line with the study goals, allowing for the approach to be used in future studies that are comparable.

- **Business rule validation**

  Businesses often employ proprietary data and procedures. These practices adhere to the company's internal principles and values and are exclusive to it. Business rule validation is the process of verifying that data is correct and consistent in accordance with certain corporate regulations. It greatly facilitates the automation of data audits and the verification of adherence to data legislation and industry best practices.

- **External data validation**

  Only data that satisfies predetermined criteria may be processed or added to the dataset when external data validation is used. A database program that only takes numerical data in a certain cell would be one example. In addition to preventing the inclusion of non-numerical data in the dataset, it may also stop mistakes in manual data input or the import of data that doesn't fit the requirements.

Automate data validation with Segment Protocols

Data correctness, consistency, and relevance may be checked in over a dozen methods, some of which can be completed manually. But in this era of big data, this becomes very challenging to scale, is expensive, and requires an excessive amount of human resources. By assisting teams in establishing data governance policies at scale, segment protocols enable organisations to automate a greater portion of the data validation process.

- **Review and resolve event violations:** Scheduled intervals, such as hourly, daily, or weekly reviews, should be used to examine event breaches. Data kinds of erroneous property values, missing needed properties, or values that do not fulfil conditional filtering are some examples of events that may be inspected by type. Resolving these infractions and taking further measures to ensure they do not occur again are both possible after they have been recognised.

- **Customize Schema controls:** Schema allows businesses to exclude attributes and features, identify which data events to allow or reject, and much more besides. It is possible to permanently remove from the dataset any occurrences that do not correspond to your personalised Tracking Plan, as well as any values or attributes that are deemed incorrect. Because of this, the validation process is accelerated and unnecessary data is saved throughout the merging process.

- **Transform bad data:** The Transformations technology from Segment allows organisations to correct inaccurate data by modifying it as it travels through the data stream. For instance, you may change the value of a property or rename a data field. Organisations may run older data as fresh to turn it into valuable data, but it cannot correct previous data.

- **Detect anomalies:** What's commonplace for one company could come as a surprise to another. Here's where anomaly detection really shines; it allows businesses to establish criteria for what constitutes data anomalies. A fresh uptick in infractions, as outlined in the Segment Tracking Plan, or an unexpected dip in business outside of normal slow times might be the cause.

5.3 Implementing Data Quality Frameworks

Organisations may benefit from data quality frameworks because they provide a systematic approach to evaluating, tracking, and improving data quality. Aspects of data quality such as correctness, completeness, consistency, timeliness, and dependability are addressed by these frameworks, which include a set of principles, standards, procedures, and tools.

Data quality frameworks allow businesses to lay out their data quality objectives, standards, and plans for achieving those objectives. If you want to create your own plan for managing data quality, you may use a data quality framework template as a guide.

There can be no room for error in data quality management. Both the systems that depend on your data and the choices you make as a company might be severely affected if your data quality is compromised.

For this reason, it is advisable to design a workable data quality strategy for the pipeline of your company.

Key Components of Data Quality Frameworks

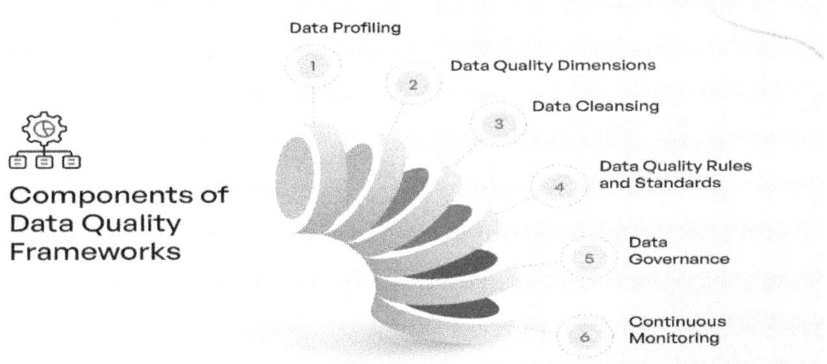

Figure 5.1: Components of Data Quality Frameworks

Source: - *(Acceldata, 2024)*

- **Data Profiling and Assessment**

  Any data quality methodology worth its salt will begin with a data profile outlining the data's features, structure, and integrity. Among these tasks is the detection of discrepancies, outliers, duplicates, and missing data. Data profiling technologies streamline this procedure while revealing data quality insights.

- **Data Quality Dimensions**

  Data quality is multi-faceted, including precision, comprehensiveness, consistency, timeliness, and applicability. Organisations can successfully assess and evaluate data quality with the help of a strong data quality framework that establishes metrics and criteria for each dimension.

- **Data Quality Rules and Standards**

  The consistency and integrity of datasets depend on the rules and standards for data quality. As a means of enforcing data quality standards company-wide, these rules specify what kinds of data values, forms, and relationships are permitted.

- **Data Cleansing and Enrichment**

  Organisations need to clean and enhance their data after identifying data quality concerns. Duplicate data removal, format standardisation, mistake correction, and missing data augmentation may all be part of this process. Automated data purification solutions make this procedure more precise and trustworthy.

- **Data Governance and Stewardship**

  Effective data asset management is the goal of data governance, which entails a set of rules, procedures, and controls. By outlining specific duties for data stewardship, a strong data governance framework promotes transparency, accountability, and conformity with applicable regulations.

- **Continuous Monitoring and Improvement**

  Ensuring data quality is a continual activity that requires constant attention. By employing continuous monitoring, organisations may identify when quality requirements are being surpassed and proactively correct them. Organisations may improve decision-making and commercial results by consistently increasing data quality.

Benefits of Data Quality Frameworks

Implementing a data quality framework offers numerous benefits to organizations:

- **Improved Decision-Making:** Making educated decisions is made possible by data insights that are accurate and dependable.

- **Regulatory Compliance:** Keeping data accurate and full guarantees compliance with industry standards and laws.

- **Operational Efficiency:** Improves operational efficiency by decreasing the amount of time and effort needed for data troubleshooting and manual interventions.

- **Enhanced Customer Experience:** Increased customer happiness and loyalty may be achieved via the use of high-quality data to create personalised and relevant experiences.

Challenges of Data Quality Frameworks

Some of the difficulties in establishing data quality standards include the following:

- **Data Silos:** Consistent data quality is difficult to achieve due to data fragmentation across systems and departments.

- **Governance Issues:** Data quality management is hindered by a lack of transparent data governance procedures and organisations.

- **Technological Complexity:** Data quality techniques and procedures need to be constantly adjusted to keep up with the ever-changing technology.

- **Cultural Resistance:** Successful implementation requires overcoming opposition to change and cultivating a data-driven culture.

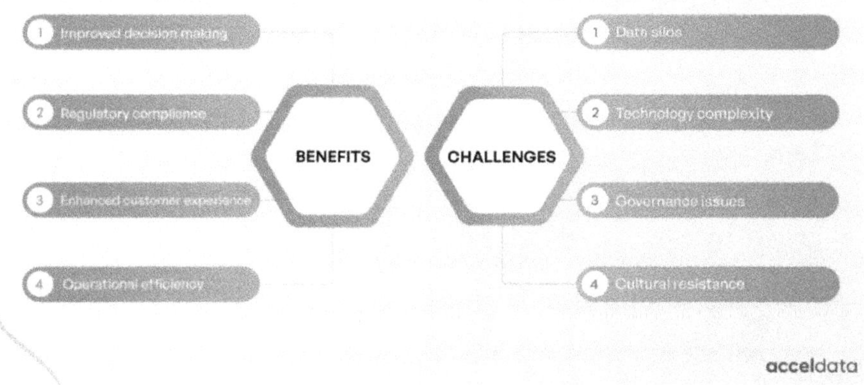

Figure 5.2: Implementing Data Quality Frameworks

Source: - *(Acceldata, 2024)*

Implementation Strategies

Implementing a data quality framework calls for meticulous preparation, implementation, and continuous monitoring. Effective implementation requires many critical tactics, such as:

- **Assessing Current State**

  There can be no room for development without first doing an exhaustive evaluation of the present data quality procedures, capacities, and obstacles.

- **Setting Clear Objectives**

  Data quality projects may be better aligned with company priorities and have demonstrable results when clear objectives, targets, and success criteria are defined.

- **Engaging Stakeholders**

  The best way to get everyone on board with data quality projects is to include everyone from the top down. This includes executives, data stewards, business users, and IT teams.

- **Investing in Tools and Technology**

  Data profiling, validation, cleaning, and monitoring may be made more efficient and successful with the use of data quality tools and technological solutions.

- **Establishing Governance Structures**

  Data quality efforts may be more reliably implemented, held to account, and maintained over time with the support of well-defined governance frameworks, procedures, and policies.

- **Providing Training and Education**

  Awareness, proficiency, and acceptance of data quality processes are improved by training and education that workers get on data quality principles, tools, and best practices.

- **Iterative Improvement**

  Organisations may continually evaluate, prioritise, and resolve data quality concerns according to changing business demands and priorities by using an iterative strategy to improving data quality.

Common Data Quality Frameworks

The most effective way to manage data quality is to refer to one of the many existing frameworks or standards that provide advice and recommendations. The following are examples of widely used frameworks:

- **TDQM (Total Data Quality Management)**

  It is an all-encompassing method that covers data gathering, data utilisation, and data quality management in general. Involving stakeholders at all levels in data quality efforts is emphasised, and the significance of integrating data quality into organisational processes and systems is highlighted.

- **DAMA DMBOK (Data Management Body of Knowledge)**

  Data quality management is one of the many data management methods that it lays forth. To help businesses execute data quality efforts, it lays out essential ideas, principles, and tasks pertaining to data quality.

- **ISO 8000**

  This one is a global benchmark for managing data quality; it specifies requirements for, and methods for, doing so. Data correctness, completeness, consistency, and integrity are just a few of the data quality issues covered. This document is useful for organisations who are looking to standardise their data quality procedures.

- **Six Sigma**

  It is a strategy for managing quality that may be used to enhance data quality by reducing process variances and faults. To make sure the data is up to par, it uses statistical methods and procedures to find and remove any mistakes.

- **Data Governance Frameworks (e.g., COBIT, ITIL)**

  Data quality is emphasised in these frameworks as a critical component of data governance, highlighting its significance in data management practices as a whole. Structures, procedures, and methods for enterprise-wide data quality management are provided by these frameworks.

Optimizing Data Quality: Implementing Data Quality Tools within a Framework

The integrity of data is of the utmost importance in the modern data-driven world. In order to keep standards high, organisations must use modern data quality technologies inside a complete framework. These technologies allow firms to detect and fix data irregularities, inconsistencies, and errors; they include data cleaning, profiling, monitoring, and validation.

A systematic framework with data quality measures, data governance, and continuous improvement initiatives is necessary for successful adoption, which goes beyond just installing technologies. A culture of data excellence can be fostered, procedures can be streamlined, and decision-making can be improved by combining these aspects smoothly. This will lead to organisational success.

A plethora of data quality technologies are available. As an example of a data quality tool, Acceldata is among the finest. The truth is that your data pipeline consists of an intricate web of different systems and technologies that transfer data from one location to another.

The integrity of your data could be jeopardised at any stage of this process. You need an observability solution that tracks your data at critical points in your data pipeline to guarantee that its quality stays good. Acceldata is specifically designed to do that very thing. You can avoid problems that impair dependability with this form of monitoring, which is a big advantage. As a result, you can guarantee constant data accessibility by removing the possibility of downtime from the equation. To get the most out of data governance, businesses may use these data quality technologies.

There are open-source data quality tools that you may come upon sometimes. There are benefits and drawbacks to these data quality technologies that you need to know. Both the UI and the functioning of open-source programs may be challenging at times. Extra time and money spent on training might be a consequence of this.

You should include instructions for maintaining good scores in each of these measures in your data quality presentation. Data quality management has a great deal of significance. You can guarantee that you are consistently making informed choices based on reliable insights by maintaining high standards of data quality.

5.4 Principles of Effective Data Visualization

It is impossible to exaggerate the significance of data visualisation to data science. The correct visualisation may help you see patterns in your data that were previously invisible, eliminate distracting noise, and bring out the most nuanced details. It's an essential tool for analysing data, testing hypotheses, and explaining results to those with and without technical backgrounds.

The Power of Visual Data Interpretation

A number of strong arguments support data visualization's status as an essential component of data science. It goes beyond just displaying numbers; it's a crucial step in the analytical process that helps stakeholders and data scientists find insights and make smart choices. The importance of visualisation may be explained as follows:

- **Immediate Insight:** Visualisations provide quick understanding of complex datasets. A well-made chart or graph may provide the same information quickly and easily as a lengthy report or spreadsheet.

- **Pattern Recognition:** Humans are visual beings by nature, and when given with visual information, our brains are programmed to swiftly identify patterns and anomalies. This is made possible by data visualisation, which draws attention to patterns, cycles, and anomalies in data that could otherwise be missed in text-based data.

- **Facilitates Communication:** Data visualisation helps simplify difficult information for both technical and non-technical audiences. Visual aids facilitate greater communication and teamwork by making it simpler to explain and debate results.

Examples Demonstrating the Impact of Visualization

Heat Maps for Weather Data: Instead of digging through tables of temperature data, you can just look at a heat map to see how the temperature has changed across various regions. This makes it much simpler to find climatic hotspots and trends.

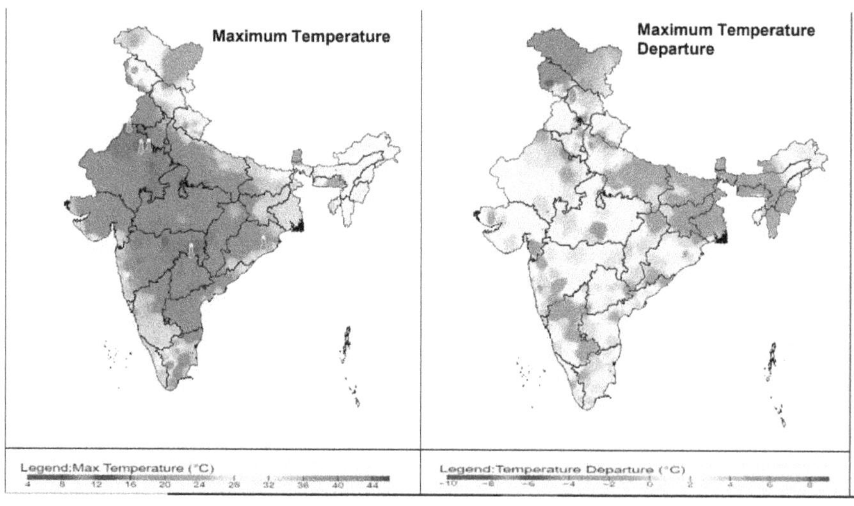

Source: - *(Shukla et al., 2017)*

Line Graphs in Financial Analysis: Stock price line graphs show historical price changes, giving investors a clear view of market patterns and allowing them to make faster investment choices.

Source: - *(Ahmed, 2024a)*

Network Diagrams in Social Science Research: Relationships and social connections may be better understood with the use of network diagrams, which can reveal community groups or influential individuals that would otherwise go unnoticed in raw data.

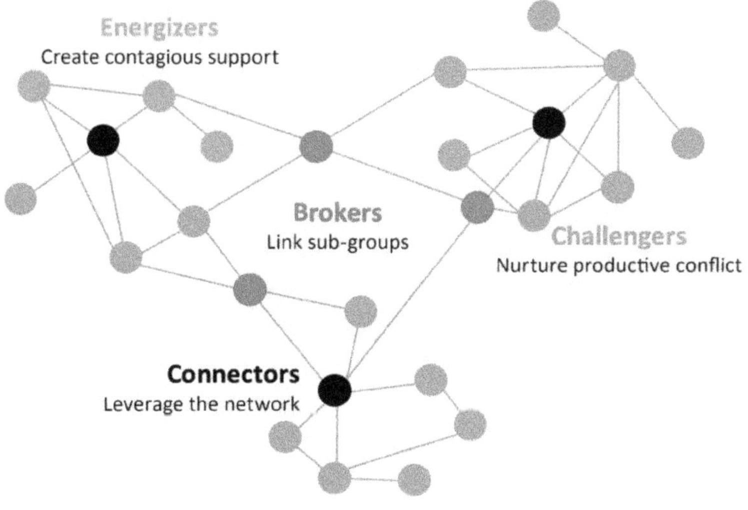

Source: - *(Ahmed, 2024a)*

Cognitive Aspects of Visual Information Processing:

- **Faster Processing:** The occipital lobe of the brain is responsible for visual processing, which makes information absorption easier and quicker than with text, which necessitates sequential processing and increases cognitive load.

- **Memory Retention:** People are more likely to recall visual content than textual information. Visually displayed information has a higher likelihood of being remembered correctly due to a phenomenon called the image superiority effect.

- **Engagement and Emotion:** Data storytelling is greatly enhanced by visuals, which have the ability to captivate audiences and elicit emotions more so than words alone. By appealing to the audience's emotions, the message may be more impactful and easier to remember.

Principle I: Clarity and Simplicity

Clarity and simplicity are not only aesthetic considerations in data visualisation; they are guiding principles that establish the visual message's efficacy. I will explain why it is crucial to prioritise these elements:

- **Enhanced Comprehension:** The visuals are simpler to interpret if they are simple and straightforward. They highlight the most crucial aspects of the data for the audience to focus on, allowing for easier and more thorough understanding.

- **Effective Communication:** Making ensuring your message gets across clearly is essential when trying to convey complicated data findings. A larger audience, regardless of their level of knowledge, can understand it because of its simplicity.

- **Increased Engagement:** Audiences are more likely to interact with simple-to-understand visualisations. Overly complex images might cause irritation and disinterest.

Tips for Achieving Clarity and Simplicity:

- **Minimize Clutter:** Get rid of everything that doesn't help with deciphering the data. Excessive use of colours, grid lines, and labels is part of this.

- **Use Clear Labels and Legends:** Labels should be legible and simple to interpret. Keep legends brief and put them next to the relevant material.

- **Limit Your Color Palette:** An excess of hues could be disorienting and unpleasant. Highlight important data points or group similar things by using colour on purpose.

- **Focus on One Main Message:** A single important idea should be conveyed by each visualisation. Think about using many charts if you have more than one message.

Examples of Clear vs. Cluttered Visualizations:

Source: - *(Ahmed, 2024a)*

Principle II: Choose the Right Chart Type

For data visualisation to be successful, choose the correct chart type is key. Selecting the appropriate graphic determines the level of clarity or confusion caused by the data's narrative. For certain data types and analyses, different styles of charts are more appropriate:

- **Bar Charts:** Perfect for comparing amounts in various areas. For broad comparisons, use vertical bars; for lengthy category names, use horizontal bars.

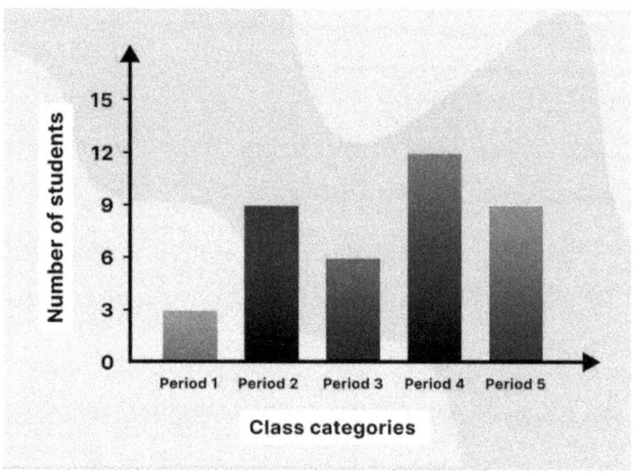

Source: - *(Ahmed, 2024a)*

- **Line Charts:** Ideal for displaying chronology. Data points that show how something's value has changed sequentially are connected by lines in the graphic.

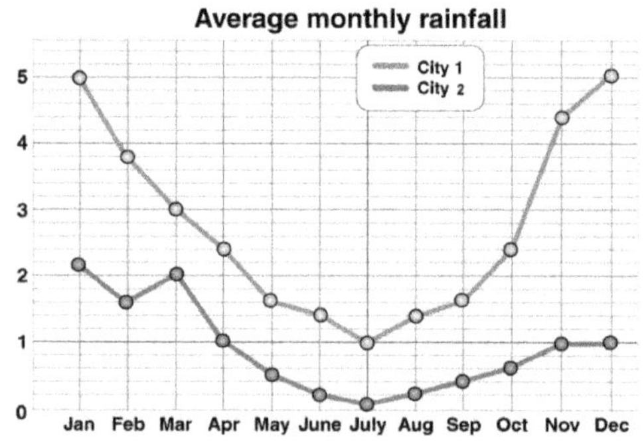

Source: - *(Ahmed, 2024a)*

- **Scatter Plots:** The best way to show the connection between two variables. Clusters or correlation patterns can be revealed by the points on the scatter plot.

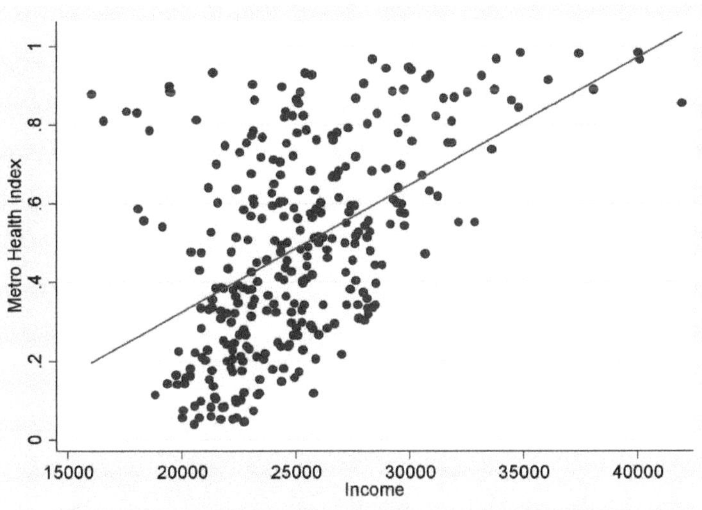

Source: - *(Ahmed, 2024a)*

- **Pie Charts:** Apt for displaying component parts. When there aren't too many categories and proportionate sizes are the main point, they work wonders.

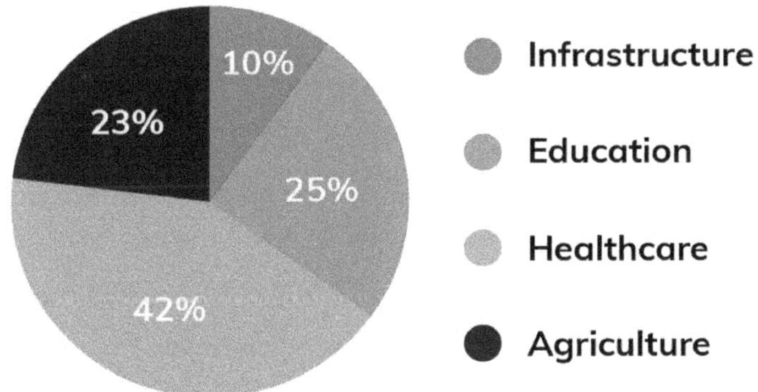

Source: - *(Ahmed, 2024a)*

- **Histograms:** Used to display the data distribution across several intervals or bins. Perfect for figuring out the distributional form of your data.

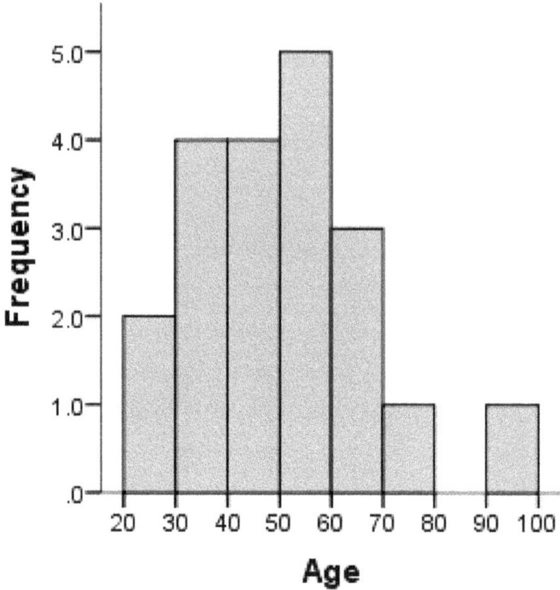

Source: - *(Ahmed, 2024a)*

Guidelines for Selecting the Right Chart Type:

1. **Identify Your Goal:** Select the appropriate data type to display: value comparisons, distributions, relationships between variables, or data composition.

2. **Consider Your Data:** The optimal data visualisation strategy will depend on the type of your data, whether it's continuous, time series, or categorical.

3. **Simplicity is Key:** Find the simplest chart that serves your purpose. People may get confused by charts that are too complicated.

4. **Follow Data Visualization Best Practices:** For accuracy and integrity, make sure your chart follows data visualisation best practices.

Principle III: Color Usage

Data visualisation that makes strategic use of colour can greatly improve the data's effectiveness by drawing attention to key points, creating an emotional impact, and differentiating the data from competing views. On the other hand, if colours aren't used correctly, they might distract from the data, which can cause misunderstandings and an ugly presentation. Colour is a powerful tool in data visualisation; here are some ways to make the most of it.

The Role of Color in Data Visualization:

- **Differentiation:** Data visualisations benefit from using colour to demarcate data sets or categories, which aids the audience in following and understanding individual data points.

- **Emphasis:** Data points, outliers, or regions of interest can be deliberately highlighted using colour to help the viewer focus on crucial findings.

- **Aesthetic Appeal:** Colour enhances visual appeal and has the potential to make data visualisations more captivating and easier to remember.

- **Emotional Influence:** It is possible to emphasise the importance, trendiness, or urgency of the data given by using colours that elicit emotions and impact perception.

Guidelines for Choosing Color Schemes:

- **Use Contrasting Colors for Clarity:**

  Different parts of the visualisation can be better seen with the use of contrast. To make text legible, you can do things like utilise contrasting colours for the backdrop and foreground.

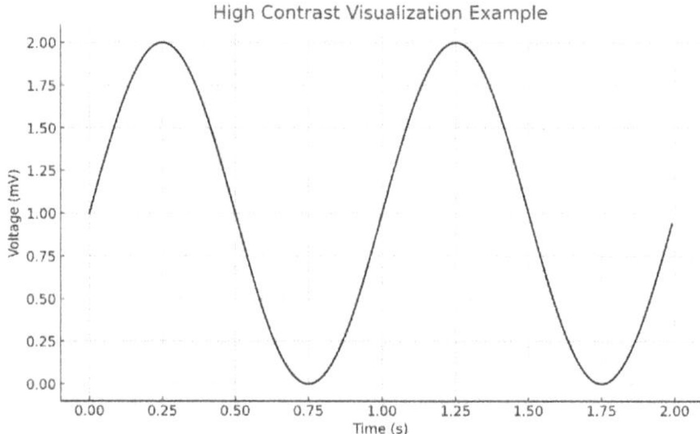

Source: - *(Ahmed, 2024a)*

Here we see a visualisation that makes use of contrasting colours to enhance clarity. The data is represented in this plot by a black line, which is a high contrast colour, set against a white background. With this option, the data will be easily readable and will stand out. The chart's basic grid with light grey lines makes it easier to read without drawing attention away from the key numbers.

- **Limit Your Palette:**

  Too many colours might be overwhelming and confusing to the eye. employ a limited colour palette, ideally 2-4 core colours, and employ shades of these colours to differentiate.

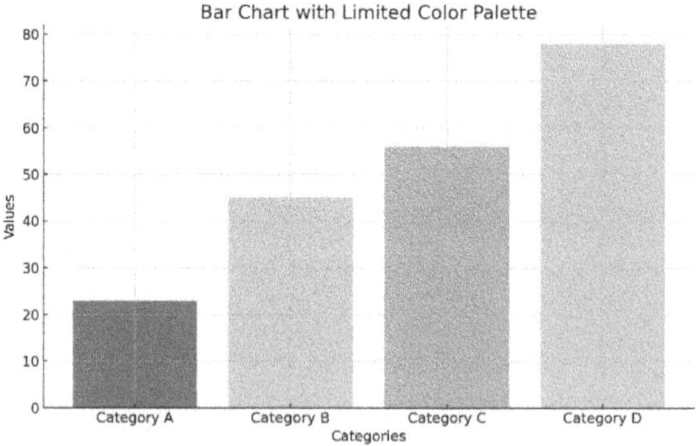

Source: - *(Ahmed, 2024a)*

This bar chart uses a small colour pallet of varying colours to depict four separate groups. This method shows how different categories can be highlighted in a distinct way while yet sticking to a consistent visual concept; it also shows how subtle colour can communicate a lot without being distracting. The use of a limited palette in data visualisation is demonstrated by the various colours for each bar, which make it easier to differentiate between groups at a glance.

- **Be Mindful of Color Associations:**

    A person's emotional reaction and cultural associations could be influenced by the colour they see. For instance, green might signify progress or security, while red is typically associated with reduction or urgency. Take into consideration cultural variations in colour perception while utilising these linkages.

- **Ensure Accessibility:**

    Make sure that anyone who have trouble seeing colours can use the colour palettes you choose. You can test how your graphics will look to people who are colourblind using tools like Colour Oracle or WebAIM.

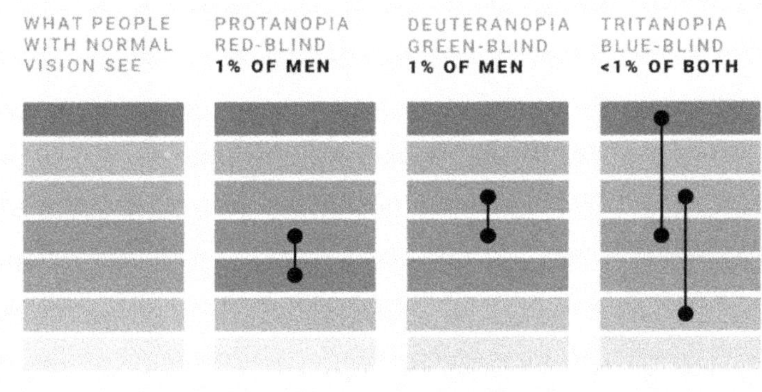

SUGGESTION FOR A COLORBLIND-SAFE PALETTE FROM THE INTERNET

Source: - *(Ahmed, 2024a)*

- **Highlight, Don't Distract:**

  To emphasise key aspects in the data, use saturated or brilliant colours; to downplay or ignore less important areas, use neutral or subdued colours. This aids in keeping the audience's attention on the most important aspects.

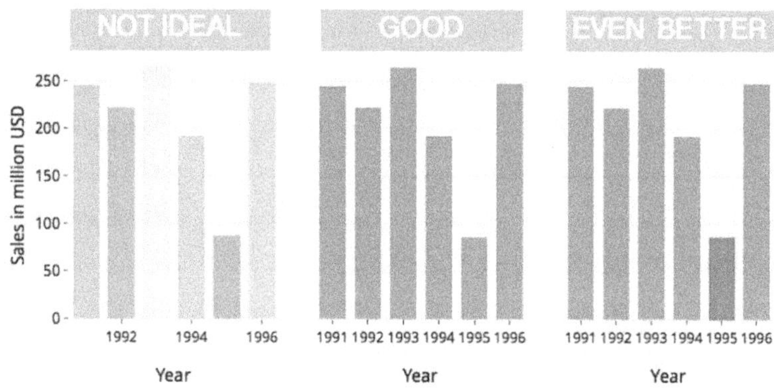

Sales in million USD from 1991–1996. Red colour is used to draw attention to unusually low sales in 1995. Nearly uniform sales in other years are all rendered in grey. [made by author]

Source: - *(Ahmed, 2024a)*

Common Color Pitfalls to Avoid:

- **Overuse of Colors:** When there are too many colours used in a visualisation, it becomes difficult to understand and interpret.

- **Ignoring Cultural Contexts:** Diverse cultures have diverse meanings for the same colour. Make sure the colours you choose don't send the wrong messages to a large audience.

- **Lack of Consistency:** The viewer may become confused if colours are used inconsistently across several visualisations. For ease of comparison and continuity, keep the colour schemes consistent throughout related charts.

Illustrating the Impact of Color Usage:

It may be challenging to tell among slices in a pie chart with too many identical colours, but distinctions can be made clearer by giving each slice a unique colour and a consistent legend.

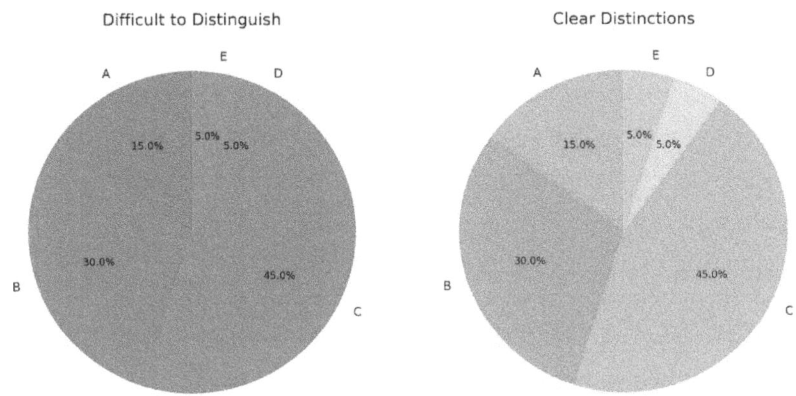

Source: - *(Ahmed, 2024a)*

Data density or concentration can be efficiently displayed with a heat map that uses a gradient of colours from chilly to warm. A heat map with bright, haphazard colours, however, may mask patterns rather than draw attention to them.

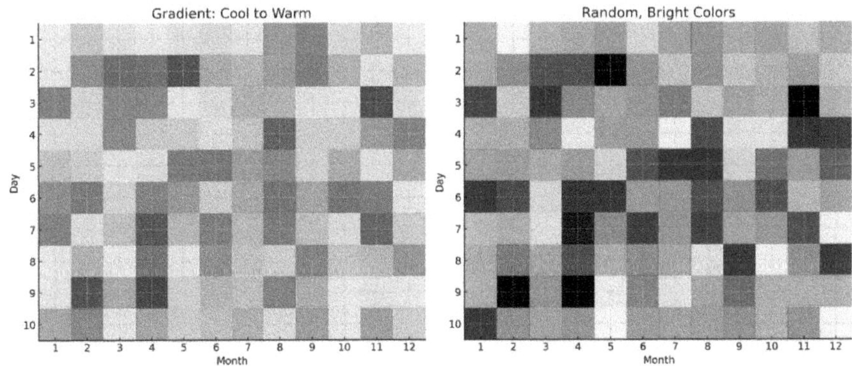

Source: - *(Ahmed, 2024a)*

Data visualisation is made more effective and easier to understand when colours are used correctly. Your data storytelling will be more effective if you follow these rules to make sure your visualisations are beautiful, easy to understand, and packed with useful information.

Principle IV: Storytelling with Data

Data storytelling is an effective method that goes beyond simple presentation by weaving together facts and figures to generate an engaging and informative story. Crucial to this process is effective data visualisation, which takes raw data and makes it into a captivating narrative with the power to motivate, educate, and reveal patterns. Using data visualisation for storytelling can be done in the following ways:

Highlighting the Narrative Power of Data Visualization:

- **Engage Your Audience:** The audience is more open to the insights offered in a data tale that is well-told. Numbers alone aren't enough to convey the significance of data; visuals may do the same.

- **Simplify Complex Concepts:** Data visualisations simplify otherwise overwhelming information for the benefit of both the audience and the information's long-term memory.

- **Reveal Trends and Patterns:** Data visualisation helps to reveal patterns and trends that were previously invisible in the raw data, providing insight into temporal trends, linkages, and causality.

Techniques for Narrating a Story Through Data Visuals:

1. **Start with a Clear Message:** Get the storyline down on paper before you start making images. Just what is the most important thing you want people to remember from what you've said?

2. **Use a Logical Sequence:** Make sure your visuals are structured in a way that logically leads the audience through the story. In order to conclude or provide a call to action, one must first set the context, then proceed with the analysis.

3. **Highlight Key Points:** Highlight the most crucial pieces of information that move the plot along by using visual features like as size, colour, or annotations.

4. **Incorporate Visual Variety:** A variety of visualisations can help illustrate various parts of your tale and maintain audience interest. When appropriate, combine graphs, charts, and pictures.

5. **Connect the Dots:** Provide context for each visualisation by using text, subtitles, or interactive features to demonstrate its role in the story. It would be a mistake to think the images convey everything (Ahmed, 2024b).

5.5 Tools for Data Visualization

Chartist, Google Charts, Tableau, Grafana, Fusion Charts, Data wrapper, In fogram, Chart Blocks, and many more are among the top data visualisation tools. These tools should be able to manage a vast amount of data, be user-friendly, and support multiple visual styles.

The significance of data is growing daily. When making key decisions, data is king, and any organisation can tell you that. Data visualisation is also attracting a lot of interest for this reason. There are state-of-the-art software programs and technologies available for data visualisation. Software for creating visual representations of data is known as a data visualisation tool. At its most fundamental, each application lets you input a dataset and make graphical changes to it. However, their features do differ. There are pre-made templates available for making basic visualisations using the most of them.

What Do the Best Data Visualization Tools Have in Common?

Every data visualisation technology that is currently on the market shares some characteristics. Their ease of use is their primary benefit. You will probably come across two kinds of software: those that are simple to use and others that make data visualisation extremely challenging. Some are built in ways that are easy to use and come with good documentation and tutorials. Some are lacking in specific areas, making them ineligible for inclusion on any list of "best" tools, regardless of other attributes. You should make sure that the software can manage a lot of data and a variety of data types on a single screen.

The best programs also have the ability to make many other kinds of graphs, charts, and maps. Others in the market will undoubtedly offer data in a slightly different way. Some data visualisation tools are very good at making just one kind of chart or map. You could say that are some of the "best" tools out there. Lastly, there are monetary considerations. A tool's bigger price tag shouldn't be seen as a red flag, but it should be accompanied by better support, more features, and overall value to justify the investment.

1. Tableau

Two key reasons contribute to Tableau's immense popularity as a data visualisation tool: its ease of use and its power. Many data sources can be connected to it, allowing you to construct a wide variety of charts and maps. Tableau is owned by Salesforce and is utilised by several individuals and large corporations.

There are various versions of Tableau, including desktop, server, and web-based versions, as well as some CRM applications.

A powerful tool for AI, ML, and Big Data applications, Tableau integrates with complex databases like Teradata, SAP, My SQL, AWS, and Hadoop to quickly generate visual representations and images from massive, dynamic information.

The Pros of Tableau:

- Excellent visualization capabilities
- Easy to use
- Top class performance

- Supports connectivity with diverse data sources

- Mobile Responsive

- Has an informative community

The Cons of Tableau:

- The pricing is a bit on the higher side

- Auto-refresh and report scheduling options are not available

2. Dundas BI

Dundas BI facilitates the development of impromptu, multi-page reports by providing highly adaptable data visualisations including interactive scorecards, maps, gauges, and charts. Dundas BI streamlines the tedious process of cleaning, examining, transforming, and modelling large datasets by giving users complete control over visual components.

The Pros of Dundas BI:

- Exceptional flexibility

- A large variety of data sources and charts

- Wide range of in-built features for extracting, displaying, and modifying data

The Cons of Dundas BI:

- 3D charts not supported

3. Power BI

Microsoft offers its user-friendly data visualisation tool, Power BI, for both on-premise and cloud deployment options. With support for numerous backend databases, Power BI stands out as a comprehensive data visualisation solution. These databases include Teradata, Salesforce, PostgreSQL, Oracle, Google Analytics, Github, Adobe Analytics, Azure, SQL Server, and Excel, among many more. Stunning visualisations and real-time insights are created by the enterprise-level solution, allowing for speedy decision-making.

The Pros of Power BI:

- No requirement for specialized tech support
- Easily integrates with existing applications
- Personalized, rich dashboard
- High-grade security
- No speed or memory constraints
- Compatible with Microsoft products

The Cons of Power BI:

- Cannot work with varied, multiple datasets

4. Zoho Reports

In just a few short minutes, you can create and share in-depth reports with the help of Zoho Reports, which is also called Zoho Analytics. This data visualisation tool combines Business Intelligence with online reporting services. Importing Big Data from popular databases and apps is also possible with the professional visualisation tool.

The Pros of Zoho Reports:

- Effortless report creation and modification
- Includes useful functionalities such as email scheduling and report sharing
- Plenty of room for data
- Prompt customer support.

The Cons of Zoho Reports:

- User training needs to be improved
- The dashboard becomes confusing when there are large volumes of data

5. Google Charts

Coded with SVG and HTML5, Google Charts is renowned for its capacity to generate visual representations of data. It is a key participant in the data visualisation industry. In addition to zooming in and out, Google Charts is incredibly cross-platform, working with everything from iOS and Android to older versions of Internet Explorer.

The Pros of Google Charts:

- User-friendly platform
- Easy to integrate data
- Visually attractive data graphs
- Compatibility with Google products.

The Cons of Google Charts:

- The export feature needs fine-tuning
- Inadequate demos on tools
- Lacks customization abilities
- Network connectivity required for visualization

6. Visual.ly

As a data visualisation tool, Visual.ly is well-known for its extensive distribution network, which shows how a project turned out. With the help of its data visualisation team, Visual.ly makes it easy to import and export data, even to other companies.

The Pros of Visual.ly:

- Top-class output quality
- Easy to produce superb graphics
- Several link opportunities

The Cons of Visual.ly:

- Few embedding options

- Showcases one point, not multiple points

- Limited scope

7. RAW

Raw Graphs, or RAW, is compatible with tab-delimited or CSV files, among others. It connects spreadsheets with data visualisation tools. Though it is a web-based application, Raw Graphs offers strong data security and features both traditional and non-standard layouts.

The Pros of RAW:

- Simple interface

- Super-fast visual feedback

- Offers a high-level platform for arranging, keeping, and reading user data

- Easy-to-use mapping feature

- Superb readability for visual graphics

- Excellent scalability option

The Cons of RAW:

- Non-availability of log scales

- Not user intuitive

8. IBM Watson

Watson Data Visualisation, named after IBM co-founder Thomas J. Watson, is an advanced data visualisation tool that employs AI and analytical components to uncover patterns and insights in structured and unstructured data. Watson, an intelligent self-service visualisation tool from IBM, uses Natural Language Processing (NLP) to walk users through the whole process of insight discovery.

The Pros of IBM Watson:

- NLP capabilities

- Offers accessibility from multiple devices

- Predictive analytics

- Self-service dashboards

The Cons of IBM Watson:

- Customer support needs improvement

- High-cost maintenance

9. Sisense

Users have access to real-time data analytics from any location using Sisense, making it one of the most versatile data visualisation tools. In order to aid decision-makers in making data-driven decisions, the top-tier visualisation tool may extract important patterns from data and summarise statistics.

The Pros of Sisense:

- Ideal for mission-critical projects involving massive datasets

- Reliable interface

- High-class customer support

- Quick upgrades

- Flexibility of seamless customization

The Cons of Sisense:

- Developing and maintaining analytic cubes can be challenging

- Does not support time formats

- Limited visualization versions

10. Plotly

Plotly is an open-source data visualisation tool that allows for complex visualisations and offers full integration with analytics-centric programming languages such as R, Python, and MATLAB. Plotly is a popular choice for teams working together on projects, and it offers both on-premise and cloud deployment options for producing and sharing interactive visual data.

The Pros of Plotly:

- Allows online editing of charts
- High-quality image export
- Highly interactive interface
- Server hosting facilitates easy sharing

The Cons of Plotly:

- Speed is a concern at times
- Free version has multiple limitations
- Various screen-flashings create confusion and distraction

11. Data Wrapper

Data Wrapper stands out among the limited number of free data visualisation tools available. Media companies love it for its built-in charting and graphical statistics capabilities on Big Data. Users may simply build maps and charts with Data Wrapper's easy and intuitive interface, and then integrate them into reports.

The Pros of Data Wrapper:

- Does not require installation for chart creation
- Ideal for beginners
- Free to use

The Cons of Data Wrapper:

- Building complex charts like Sankey is a problem

- Security is an issue as it is an open-source tool

12. High charts

High charts, used by 72 of the top 100 corporations in the world, is an excellent tool for visualising insights on streaming big data. High charts is an interactive visualisation tool that runs on the JavaScript API and integrates with jQuery. It supports cross-browser functionality and makes it easy to view visualisations.

The Pros of High charts:

- State-of-the-art customization options

- Visually appealing graphics

- Multiple chart layouts

- Simple and flexible

The Cons of High charts:

- Not ideal for small organizations

13. Fusion charts

When it comes to data visualisation, Fusion charts is among the most used and well-known options. Users have a lot of leeway with the top-tier visualisation tool, which is built on JavaScript and offers ninety distinct chart building packages that interact with various frameworks and platforms.

The Pros of Fusion charts:

- Customized for specific implementations

- Outstanding helpdesk support

- Active community

The Cons of Fusion charts:

- An expensive data visualization solution

- Complex set-up

- Old-fashioned interface

14. JupyteR

A highly regarded data visualisation tool, JupyteR is a web program that lets users combine visuals, mathematics, narrative text, and live code into one document. Statistics, numerical simulation, data transformation and purification, interactive computing, and ML are all areas where JupyteR shines.

The Pros of JupyteR:

- Rapid prototyping

- Visually appealing results

- Facilitates easy sharing of data insights

The Cons of JupyteR:

- Tough to collaborate

- At times code reviewing becomes complicated

15. QlikView

QlikView serves more than 40,000 customers in 100 countries and is a leading data visualisation provider. Analytics, enterprise reporting, and Business Intelligence capabilities are just a few of the robust features included in Qlikview's data visualisation tool, which also allows for rapid and customised visualisations.

The Pros of QlikView:

- User-friendly interface

- Appealing, colorful visualizations

- Trouble-free maintenance

- A cost-effective solution

The Cons of QlikView:

- RAM limitations

- Poor customer support

- Does not include the 'drag and drop' feature

16. In fogram

One of the most widely used software applications available online right now is In fogram. It is an online application for data visualisation and infographic creation. Its main goal is to let all users quickly and easily create engaging dashboards, infographics, and reports using data-driven information and eye-catching visuals. Among the features offered by this specific solution are a drag-and-drop editor, 20 pre-made design templates, a large number of images and icons, and more than 550 maps and 35 charts. This tool is easy to use, even for those who are new to the field.

With its easy-to-use editor, customers may change the look and feel of their visualisations, insert company logos, and manipulate the display options. Users will also have the ability to incorporate more than a million images, GIFs, and icons into their visualisations. Users can attract traffic to their website by adding connections to interactive charts. Audiences can study data using In fogram tabs. Metrics can be set up to track how engaged an audience is via reports that are both shareable and interactive.

17. Chart Blocks

Chart Blocks handles the entire import process and chooses the right data section to make a chart. Data can be imported from almost any source. It improves a plethora of sharing options that include placing the chart on a website and immediately sharing it. There are a plethora of options for personalisation and layout that affect different parts of the graph. Using a simple chart design wizard, the Wizard feature chooses and chooses the right data for the chart. Quickly import data from any source with Chart Blocks' data import capabilities. It facilitates the generation of the chart and the import of accurate data from the intended source. This entire process only takes a few minutes. No coding is required to generate a chart.

It has a chart creator, lets you choose from hundreds of customisable chart types, and lets you create a chart in minutes. Additionally, it is capable of collecting data from almost any source and utilising it to create visualisations. You can follow the data import wizard's instructions to the letter. You can distribute and embed charts into any website you choose with ease.

The integrated social media sharing features offer identical sharing capabilities. Twitter and Facebook are among the recognised services it communicates with directly. It also has a feature that lets you export the charts as images and vectors that you can alter.

18. D3.js

Data-Driven Document enables remote document manipulation by offering data binding to DOM elements in any browser. Data transformation requires picking and choosing which nodes to work with and then doing so on an individual basis. The document's underlying DOM can be accessed by working with data functions, which allow you to change and alter node attributes, register event listeners, change nodes, modify HTML or text content, and more. To enhance performance, you can link nodes with activities like adding, updating, and deleting. The bundled graphical primitives and function factory allow you to construct new functions with ease. A function, rather than a constant, can be used to retrieve geographic coordinates. Data binding allows for the reuse of properties in documents.

For example, it can take data and generate an HTML table using CSS, SVG, and HTML. You can support massive amounts of data, see it visually in bar charts and graphics, and have fun interacting and animating in a 3D environment with all that data thanks to the fast performance and animated transitions.

19. Chart.js

An open-source charting toolkit for JavaScript, Chart.js is widely used. You can use it to create visual representations of data; it's software for data visualisation. Since it is open-source, everyone can contribute to its upkeep. Eight different kinds of charts, such as pie, line, and bar charts, are compatible with it. The bright side is that they are all incredibly responsive. Simply upload your chart, and the library will ensure its legibility. Wrappers for several frameworks are available in its robust ecosystem, which has garnered 53.7k ratings on GitHub. The charts are rendered on the canvas of the browser by the library. Many members of the community have contributed to this autonomous initiative. In addition to bar charts, bubble charts, scatter plots, line graphs, and polar charts are also available in Chart.js.

20. Grafana

Grafana is free and open-source analytics and visualisation application. It allows you to query, view, monitor, and analyse logs, traces, and metrics that are saved everywhere. It contains tools for creating educational graphs and visualisations using time-series database (TSDB) data.

The Graohana cloud is another feature. This Open SaaS logging and analytics platform is fast, fully managed, and highly accessible. The application offers every feature you enjoy about Grafana, but it is hosted and managed by Grafana Labs.

Grafana Enterprise is the paid version of Grafana that includes features that aren't available in the free and open-source version. With Grafana Corporate, you get access to business data sources, advanced authentication options, granular permission controls, round-the-clock support, and training for your core team.

21. Chartist.js

Create a library or libraries of high-quality, fully-customizable, responsive charts with the help of the web tool Chartist.js. So that it may be used in an approachable framework, Chartist.js stores the provided data in a library. Many projects now use Chartist.js to create libraries. Some examples include Chartist JSF (Java Server Faces Component), node chartist (node package for server-side charts), ng-chartist.js (Angular Directive), Table press Chartist (WordPress/table press extension), Ember - cli - chartist (Ember Addon), react chartist (react component), and many more.

The fact that Chartist.js works with so many different browsers makes it very user-friendly. Many impressive features are made possible by the browsers. These include cross-browser compatibility, advanced CSS and SVG animations, responsive option override, multi-line labels using SMIL, and more.

In order to deliver trustworthy data, these are essential qualities that each browser should strive to achieve. The ability to add motion to charts is a feature of Chartist.js that makes them more appealing and easier to read.

22. Sigma.js

One library that may be used to draw graphs with JavaScript is Sigma. Making it easy to deploy networks on websites, it also lets developers integrate network exploration into complex web apps.

- The Sigma.js layout is fantastic.
- It enables individuals to follow up with interest as soon as possible.
- Sigma.js's performance is currently satisfactory.
- Sigma.js support is fantastic and quite helpful.
- Good software must be tried.

23. Polymaps

Polymaps is an open-source JavaScript library for SVG-based image and vector tiled maps that was developed by SimpleGeo and Stamen. The library facilitates the development of dynamic and interactive online maps, the quick presentation of datasets, and the support for various visual representations of vector data (tiled). Polymaps enables cartography from several image-based web map providers, including Cloud Made, OpenStreetMap, Bing, and the like. It is capable of loading data at any scale and effectively displays information at any level, from the national to the municipal. It uses Scalable Vector Graphics (SVG) for data presentation, which makes data design easy with CSS rules. Users also don't have to learn new scripts because they can utilise the ones they already know for most jobs(Kumar, 2024).

5.6 Crafting Meaningful Reports

Data reporting has become more important for organisations throughout the world as a result of the digital age and the deluge of data it has generated. When it comes to defining company strategies, making operational choices, and tracking performance, effective data reporting is crucial. Nevertheless, it can be difficult to make sense of and display large volumes of data in a way that is both valuable and understandable. Key concepts for effective data reporting are outlined in this article (Briney, 2015)university, publisher, and funder requirements for data management by researchers changed rapidly. Government open-data mandates, data management plans, and the development of institutional and disciplinary data repositories have put an end to casual storage of data on graduate assistants' laptops. In Data Management for Researchers: Organize, maintain and share your data for research success, Kristin Briney, a Data Services Librarian at the University of Wisconsin–Milwaukee with research experience in Chemistry, provides a brief introduction to data management. This practical handbook can help bring new researchers quickly up-to-speed on the topic, as well as serve as a reference to meet specific data management needs they encounter throughout the data life cycle.",author":[{"dropping-particle":"","family":"Briney","given":"Kristin","non-dropping-particle":"","parse-names":false,"suffix":""}],"container-title":"Journal of eScience Librarianship","id":"ITEM-1","issue":"2","issued":{"date-parts":[["2015"]]},"page":"1-4","title":"Data Management for Researchers: Organize, maintain and share your data for research success","type":"article-journal","volume":"4"},"uris":["http://www.mendeley.com/documents/?uuid=cbecbffa-9c70-4679-bea5-4eac51617256"]}],"mendeley":{"formattedCitation":" (Briney, 2015.

1. Define Clear Objectives

Make sure you know what you want out of the data reporting process before you begin. In writing this report, what are you hoping to accomplish? May I ask you which questions? To what extent do you wish to influence decision-making? You can better plan your data gathering, analysis, and reporting if you answer these questions.

2. Collect and Use Relevant Data

Your report will not benefit from all data. Before you can go on to further data, you must determine which data is pertinent to your goals. Time and resources are wasted and results might be skewed when unnecessary data is collected. Also, make sure the data you use is accurate at all times. Using erroneous data can result in poor insights and actions that could harm a company.

3. Create a Structured and Consistent Format

To effectively report data, a format that is both structured and consistent is required. This will make sure that your reports are simple to read and compare with other reports. Title, introduction, body (including data and insights), and conclusion/recommendations are the essential components of any good structure. All of your reports should also make consistent use of the same units, terminology, and definitions.

4. Use Visuals to Represent Data

The use of images in reports can substantially enhance comprehension, as humans are primarily visual beings. Data that is difficult to grasp can be simplified and presented more clearly with the use of charts, graphs, and diagrams. Just make sure you pick the correct image to go with your data. Visualise percentages with a pie chart, trends over time with a line graph, and categories with a bar chart.

5. Simplify Your Reporting

When reporting data, simplicity is crucial. Stay away from jargon and technical phrases that your readers might not be familiar with. Regardless of the level of technical expertise in your audience, you should strive to ensure that your reports are as easy to read as possible. It is important to provide clear explanations when using technical words.

6. Make Your Reports Actionable

Insights that can guide action should be included in your data reports, not merely facts. The facts offered in each report should be accompanied by practical suggestions for next steps. By doing so, you ensure that your audience comprehends both the data's meaning and the actions they can take in response.

7. Validate and Review Your Reports

Be careful to double-check your data and analysis before submitting your reports. Verify that your data is reliable and free of mistakes; furthermore, make sure that your analysis makes sense. You can keep the integrity of your reports and avoid embarrassing situations by following this step.

8. Keep Data Secure

The safety of reported data is of the utmost importance. To prevent data loss, alteration, or theft, you should implement robust security procedures. Safeguards include limiting who can access data, storing and transmitting data securely, and auditing and monitoring data operations on a regular basis.

Multiple Choice Questions (MCQs)

1. **Which of the following is a key data quality metric used to assess the accuracy of data?**

   a. Timeliness

   b. Completeness

   c. Validity

   d. Uniqueness

2. **Which method of data validation ensures that the data adheres to predefined formats or rules (e.g., dates in "YYYY-MM-DD" format)?**

   a. Range checks

   b. Format checks

   c. Consistency checks

   d. Uniqueness checks

3. **Which of the following is a common framework used for ensuring data quality in an organization?**

   a. Six Sigma

   b. Data Governance Framework

   c. Lean Methodology

   d. Agile Framework

4. **What is the primary purpose of data visualization?**

   a. To summarize large datasets into a single number

   b. To display data in a way that highlights key trends and patterns

   c. To store data efficiently

   d. To encrypt sensitive data

5. **Which of the following is a popular tool for creating interactive data visualizations?**

    a. Microsoft Word

    b. Tableau

    c. SQL Server

    d. Python (NumPy)

6. **When crafting a meaningful report, which of the following is an important aspect to consider?**

    a. Providing a detailed technical description of the data source

    b. Including only raw data without any analysis

    c. Tailoring the report to the needs of the audience

    d. Using complex jargon to demonstrate technical expertise

7. **Which of the following data quality metrics refers to the degree to which data is free from errors or mistakes?**

    a. Accuracy

    b. Timeliness

    c. Completeness

    d. Consistency

8. **Which validation method involves comparing data across multiple sources to ensure consistency?**

    a. Range checks

    b. Consistency checks

    c. Format checks

    d. Uniqueness checks

9. **Which of the following is a key principle of effective data visualization?**

   a. Overloading the visualization with as much data as possible

   b. Using multiple charts to represent the same data

   c. Keeping visualizations simple and focused on key insights

   d. Avoiding the use of colors or visual distinctions

10. **Which tool is widely used for creating static data visualizations using code, typically in a Python environment?**

   a. Power BI

   b. Excel

   c. Matplotlib

   d. QlikView

Answer

1	2	3	4	5	6	7	8	9	10
C	B	B	B	B	C	A	B	C	C

Chapter 06

DATA SECURITY AND PRIVACY

6.1 Data Encryption Techniques

The process of converting information from a readable form, called plaintext, into an unreadable one, called ciphertext, is the essence of encryption. This is an essential procedure for protecting sensitive data from prying eyes, particularly when sent over the internet. A text-cyphering mechanism known as a "Scytale" originated with the ancient Greeks.

Encryption is constantly in motion whenever we engage in online activities like as making a purchase, sending an email, or even simply browsing the web. Anyone utilising digital technology can benefit from understanding it, but it is especially important for cybersecurity specialists and IT experts. When it comes to how it works, encryption involves applying several algorithms and encryption keys to the plaintext. The only way to decipher the generated ciphertext into plaintext is to have the appropriate decryption keys.

Common Types of Encryption

While there are many various kinds of encryption, the three most popular and extensively utilised are hashing, symmetric encryption, and asymmetric encryption. Now, let's examine each method in detail.

1. **Symmetric Encryption**: An old and popular one is symmetric encryption. The basic idea behind Symmetric Encryption, which is sometimes called Secret Key Encryption, is to use a single key for both encryption and decryption. The key, which is usually a computer-generated random string of bits, can be used for either method. It's easy to picture this key as a detailed blueprint for a coded language. For example, switching out "A" for "Z", "B" for "Y", and continuing in this manner until the entire alphabet is reversed. Using this secret language, you flip each letter according to your key when you write a letter (plaintext). Transforming a legible message into an unintelligible muddle is your encryption method. However, if a buddy of yours learns your secret language and has the key, they can reverse the order of the letters.

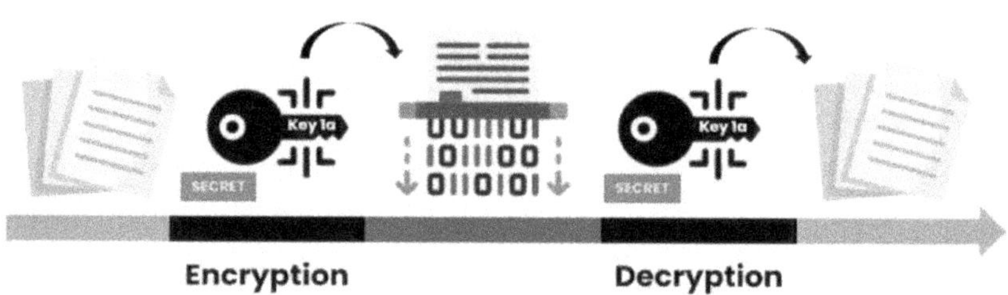

Sources: - *(Sealpath, 2023)*

Examples: A locker service is a simple example that everybody can comprehend. Put your valuables in a locker and secure it with the key. To access the locker and get your belongings, you'll need that identical key. Symmetric Encryption's strength and simplicity are exemplified by this. Symmetric Encryption is a part of our everyday lives; it protects commonplace activities like watching satellite TV, banking online, and surfing the web.

Pros and Cons

Key secrecy is fundamental to the strength of encryption, even if it is simple and fast (faster than other forms). Data confidentiality requires secure transmission of the decryption key if it must be shared. When dealing with several users or remote connections, Symmetric Encryption is not the best choice because your data safety is at risk if the key is compromised. In summary, Symmetric Encryption is a mainstay in the encryption field due to its mix of nimble speed and simplicity, which offers inherent value despite its limits.

2. **Asymmetric Encryption:** Asymmetric Encryption makes use of two keys, in contrast to Symmetric Encryption's use of a single key for both processes. There is a public key for encryption and a private key for decryption. Picture a door that requires two keys: one to lock it (the public key) and another to open it (the private key). Turning the key to the right, pressing the door handle, and pushing the door shut can be the steps needed to lock the door. I can do this. Anyone can. The key must be inserted, turned to the left, and then pulled in order to unlock it. This action can only be executed by you, the keyholder. Two linked but separate keys are formed via advanced mathematical functions. Algorithms like RSA and ECC create a public key and a private key in such a way that only the private key can decode a message encrypted with the public key.

Sources: - *(Sealpath, 2023)*

Examples: Take the mailbox for instance. It provides a simple example that everybody may grasp. The public key allows anyone to put a letter in the mailbox, but only the private key holder may open the mailbox and collect the messages. Many applications make use of asymmetric encryption or public key cryptography; one example is Secure Web browsing, which is denoted by the "https" in Secure HyperText Transfer

Protocol Secure. Email security is an additional area of use. To safeguard email communications, technologies such as "Pretty Good Privacy" (PGP) and "S/MIME" (Secure/Multipurpose Internet Mail Extensions) utilise asymmetric encryption.

Pros and Cons

Anyone can transmit encrypted messages to the owner of the key using Asymmetric Encryption because the public key can be freely circulated. An inventive solution to the key exchange problem in Symmetric Encryption is that only the recipient possesses the private key, which allows them to decipher these messages. Its use has been critical in solving the difficult key distribution problem in Symmetric Encryption and ensuring the security of communications between peers in a network.

Asymmetric encryption is slower than symmetric encryption because key creation and data transformation require considerable calculations. Here, Symmetric Encryption comes into play; it's commonly employed for transmitting data, while Asymmetric Encryption safeguards the transmission of the symmetric key. Despite requiring more calculations, Asymmetric Encryption is crucial in the safe digital realm due to its inherent secure key exchange value and its essential involvement in numerous encryption algorithms.

3. **Hashing:** The term "hashing" refers to a special method of converting any data, regardless of its size, into an output with a fixed length and unique identity: a hash value. Imagine this procedure as the making of a special label for a book. All scripts, regardless of length or subject matter, are assigned a unique ISBN for easy identification. The output (hash) of a hashing algorithm is different for every single piece of input data. It takes an input, be it an email with only one word or a whole library of ebooks, and generates a unique hash by running it through a series of complicated computations. It bears repeating that a hash is completely independent of the input, down to the slightest detail, such as a single character change.

Sources: - *(Sealpath, 2023)*

Examples: Multiple applications rely on hashing, including digital signatures, data integrity checks, password storage, and data retrieval. You might recognise it from its common use in search engines. You can see this in action when you conduct an online book search using the ISBN; the system employs a hashing algorithm to swiftly find and get the information. Each time you set a password for a website, it is kept and hashed. A new hash of the password is generated and compared to the one stored each time you log in. The fact that the real password is never saved is a huge plus for safe password management.

Pros and Cons

The fact that hashing is 'one-way' is an important feature. Although hashes can be generated from inputs, the original data cannot be recovered from them. Neither the recipe card nor the hash can be used to determine the exact contents of a dish, nor can it be used to restore the original input. But hashing isn't without its difficulties. When two separate inputs produce the same hash output, we say that there has been a hash collision. Due to the vastness of a decent hashing algorithm's outputs, this is very rare to happen, but it is still a possibility and a concern in hashing.

Hashing is vital in the digital world despite these obstacles since it promises data integrity, rapid data retrieval, and secure password storage. Its stealthy yet powerful activities keep our digital interactions secure and legitimate by reliably guarding data integrity.

The process of data encryption involves converting the data into ciphertext, which can only be deciphered using a special key that was generated either before or at the moment of encryption. This ensures that the data remains confidential. Encryption is the process of changing plaintext into ciphertext.

Key Objective of Encryption Data

- **Confidentiality:** Data encryption makes guarantee that no one other than authorised users may read or understand the data.

- **Data Integrity:** The fact that encrypted data does not change while in transit is another way encryption ensures data integrity. Encrypted data will fail integrity checks or become unintelligible if altered by unauthorised parties.

- **Authentication:** To ensure the identity of the sender or receiver of a message, authentication procedures may include encryption.

- **Non-Repudiation:** Encryption allows events to guarantee that they cannot deny contributing to the growth or transmission of a certain data item.

Importance of Data Encryption

Encryption is crucial and should not be underestimated. The fact that your data is kept on a common infrastructure does not mean it is immune to hacking. While data breaches can and do occur, encrypting your data can significantly reduce the likelihood of such an occurrence. Take a second to think about it in this light. Encrypting data before delivering it ensures its safety, even if the system it is stored on is secure. Protectiveness is lower in sanctioned systems.

Imagine for a second that the user is able to view confidential information as they work. In the absence of encryption, the user is free to save the data anywhere they like on a removable disc. Users can still make copies of data if encryptions are pre-set, but the data will be unreadable when viewed elsewhere. The true worth of data encryption is demonstrated by these advantages.

How does Encryption work?

The public internet is made up of a vast network of interconnected devices all over the world that facilitate the transfer of data and information. There is always the chance that hackers may steal or hack data when it is transferred over the open internet. Users can circumvent this by installing software or hardware that ensures data transfers are secure. These procedures are known as encryption in the field of network security. Encryption is the process of changing plaintext into ciphertext. When encrypting data, a cryptographic key—a set of mutually accepted mathematical values—is utilised. The receiver uses the key to decrypt the data, making it readable again. The more complicated the cryptographic key, the more secure the encryption will be since third parties are less likely to utilise brute force attacks, which entail testing random numbers until the correct combination is found, to decipher data. Encryption also protects passwords. Password encryption algorithms randomly mix up your password, making it impossible for hackers to decode.

States of Data Encryption

- **Data encryption in transit:** Data in transit describes information that is now in motion from one location to another, whether it's through a public or private network. Weak transfer methods make data less secure while in transit.

- **Encryption of data at rest:** The longer it takes to access information and the longer it takes to discover data loss, ransomware attacks, remotely erased data, or changed credentials, the less likely it is that data will be compromised due to events like lost or stolen devices, accidental password sharing, or accidental permission granting.

Uses of Data Encryption

- Data integrity and authenticity can be proven by encryption and digital signatures. Encryption is essential for copy protection and digital rights management.

- Data can be erased via encryption. If you encrypt your data before discarding the key, no one will be able to decipher it, but they will be able to recover the ciphertext, not the actual data, even if data recovery software can occasionally restore erased files.

- When moving sensitive information over a network, it is best to use data migration to make sure that no one else can decipher it.

- Encrypt all of your data before storing it in the cloud; VPNs (Virtual Private Networks) use encryption. Not only that, but it can encrypt not only voice calls but the entire hard drive.

Advantages of Data Encryption

- Encryption creates a barrier between the data and the security of the storage device. Administrators can store and transmit data via unsafe routes with the use of encryption, which ensures security.

- The encrypted file cannot be opened if the user loses the key or password. On the flip side, data becomes unsafe and accessible to everyone at any moment when using simpler keys for encryption.

- Encryption makes our data more secure.

Disadvantages of Data Encryption

- The encrypted file cannot be opened if the user loses the key or password. On the flip side, data becomes unsafe and accessible to everyone at any moment when using simpler keys for encryption.

- Encrypting data is a worthwhile data security measure, but it requires a lot of resources including computer power, time, and a plethora of techniques for both encryption and decryption. Consequently, the strategy is not without its expenses.

- When layered with modern systems and apps, data protection solutions may be challenging to use. The regular functioning of the device can be negatively affected by this.

- Establishing unreasonable expectations and standards that could compromise data encryption protection is conceivable if companies do not take encryption systems' limitations into account.

Data Encryption Algorithms

Multiple data encryption techniques exist, each tailored to a specific use case; nonetheless, the following are among the most popular:

- An outdated symmetric encryption algorithm, DES (Data Encryption Standard) is not fit for purpose in today's world. Consequently, other encryption methods have surpassed DES.

- Triple DES (3DES or TDES): Encrypts, decrypts, and encrypts again to obtain a greater key length by repeating the DES algorithm three times. If you want it extra secure, you can run it with three keys instead of just one. Because it is a block cypher, 3DES is susceptible to attacks like block collisions.

- One of the earliest public-key algorithms, RSA is asymmetric encryption that only works in one direction. Internet users love and rely on RSA due to its lengthy key length. It is a component of numerous security protocols, including SSH, OpenPGP, S/MIME, and SSL/TLS, and it is utilised by browsers to establish encrypted connections over unsecured networks.

- Two fish is a lightning-fast method that uses a sophisticated key structure for extra security and has sizes of 128, 196, and 256 bits. It is part of the OpenPGP standard and is freely available; it is also incorporated in several top free programs, including as VeraCrypt, Pea Zip, and KeePass.

- Elliptic Curve Cryptography (ECC) is an improvement over RSA that allows for far shorter key lengths while yet providing superior security. One asymmetric mechanism in the SSL/TLS protocol is ECC.

- The United States government uses the Advanced Encryption Standard (AES) as its encryption standard. The Advanced Encryption

Standard (AES) uses block cypher techniques and is symmetric-key based. The number of encryption rounds grows as the size rises; it is available in 128, 192, and 256-bit formats. It's hardware and software implementations were intentionally kept basic.

6.2 Access Control Mechanisms

In order to safeguard their systems and confidential information, developers must have a thorough understanding of access control mechanisms. A system's resources can be accessed and managed using various processes and techniques known as access control mechanisms. It entails governing the circumstances under which certain resources can be accessed by certain individuals.

Developers may reduce the likelihood of data breaches, stop unauthorised users from gaining access, and enforce security regulations with the help of Access Control Mechanisms. Authentication, authorisation, and distinct forms of access control like RBAC, DAC, and MAC are all examples of possible procedures. Learn more about models for access control here.

What are Access Control Mechanisms?

A system's access control mechanisms are its methods and procedures for managing and controlling who has access to what resources. Protecting private information and data is their primary responsibility.

The purpose of access control mechanisms is to provide the rules for granting and restraining access to resources. Their purpose is to safeguard information, prohibit unauthorised access, and implement security regulations.

Access Control Mechanisms come in a variety of forms, each tailored to meet the unique demands and safety standards of a given system. A few examples of frequent types are:

- **Role-Based Access Control (RBAC):** This system determines which users have certain levels of access depending on their job functions. It restricts access to resources so that people can only do their jobs.

- **Discretionary Access Control (DAC):** It is possible for resource owners to manage who has access to their resources by using DAC to set access permissions. It offers a great deal of adaptability, but it also demands vigilant oversight.

- **Mandatory Access Control (MAC):** Rules and labels are used by MAC to enforce access policies. Wherever the security of sensitive information is paramount, such as in government or military systems, this is the protocol of choice.

Authentication and authorisation procedures are used in conjunction when access control mechanisms are being implemented. Authorisation establishes what a user can do and what resources they have access to, whereas authentication confirms the user's identity.

Reduce the likelihood of data breaches and illegal access by limiting access to sensitive information to authorised individuals exclusively with the help of Access Control Mechanisms. Maintaining a strong security posture requires routinely reviewing and updating access policies, implementing multi-factor authentication, and monitoring and auditing access control systems.

Types of Access Control Mechanisms

To control and manage who has access to what resources in a system, administrators can use one of many different kinds of access control mechanisms. Different types are better suited to meet specific security needs due to their distinct properties.

- **Role-Based Access Control (RBAC):** RBAC uses established roles and responsibilities to distribute access rights. Access is allowed to users according to their duties, making sure they can only access the resources they need to do their jobs.

- **Discretionary Access Control (DAC):** Resource owners can manage who has access to their resources using DAC. The owners of the resources have the power to decide who can access them and to what extent.

- **Mandatory Access Control (MAC):** MAC uses labels and specified rules to enforce access policies. When data secrecy is of the utmost importance, it is frequently employed in highly secure settings.

- **Access Control Lists (ACLs):** ACLs are a compilation of relevant permissions for a given resource. Both the types of access and the persons or groups that have them are defined by these.

- **Attribute-Based Access Control (ABAC):** Access permissions are determined by ABAC using characteristics. When making judgements about access control, it takes a number of factors into account, including user roles, environmental conditions, and resource properties.

System characteristics and individual security needs dictate the best Access Control Mechanism to implement. When selecting an access control method, it is crucial to think about aspects like scalability, flexibility, and management ease.

Role-Based Access Control (RBAC)

The Access Control Mechanism known as Role-Based Access Control (RBAC) is quite popular since it uses predetermined roles and responsibilities to determine who has authorisation to do what. Any business, no matter how big or little, can benefit from its adaptable and scalable access control solution.

Roles within an organisation are linked with access rights in RBAC. Then, according to the duties they do, users are given one or more roles. By letting administrators designate permissions to roles instead of specific people, this makes access control management much easier.

RBAC offers several benefits:

- **Simplified Administration:** Managing user rights is a tedious administrative task, but RBAC makes it easier. Assigning or deleting responsibilities makes granting or revoking access a breeze.

- **Granular Access Control:** RBAC enables granular management of access rights. User access to resources can be carefully controlled by assigning them to specific roles based on their level of privilege.

- **Scalability:** RBAC is ideal for companies that go through expansions or changes in their personnel because of how scalable it is. You can add or edit new roles without changing the rights for current ones.

- **Security and Compliance:** RBAC is useful for making sure that security policies are followed and that regulations are met. Restrict Access Control (RBAC) lessens the likelihood of data breaches and unauthorised access by allocating permissions according to responsibilities.

Roles must be defined, permissions must be linked to roles, and users must be assigned to roles in order to implement RBAC. To keep access rights in line with organisational requirements, it's critical to periodically evaluate and update role assignments.

Discretionary Access Control (DAC)

An Access authority Mechanism that gives resource owners authority over who can access what is known as DAC. The owners of the resources in DAC can set the permissions and levels of access for those resources.

When using DAC, the resource owner has complete choice over who has access to what. In addition to having the power to give or deny access, the owner also has the ability to transfer this right to other users.

Key features of DAC include:

- **Flexibility:** DAC offers a great deal of leeway since it lets resource owners to use their own discretion and standards when deciding how to manage access.

- **Ownership-based Access:** DAC is predicated on the idea of ownership, wherein the one in possession of a resource possesses the power to regulate who has access to it.

- **Access Control Lists (ACLs):** Access Control Lists (ACLs) are frequently used by DAC to define access permissions for specific resources. A list that assigns particular access privileges to people or groups is called an ACL.

While DAC is adaptable, it can be difficult to centrally manage and ensure that all users have consistent access controls. The correct assignment and revocation of access permissions necessitates meticulous preparation and organisation.

DAC is frequently employed in settings where users require autonomy over their resources, including file systems, collaboration platforms, and personal computers. Environments requiring strict regulation, centralised control, and uniform access regulations should not be used.

In order to implement DAC, it is important to determine who owns what resources, create access control lists, and train resource owners to properly manage rights.

Mandatory Access Control (MAC)

The Access Control Mechanism known as Mandatory Access Control (MAC) uses labels and predetermined rules to enforce access policies. It finds widespread application in government and military systems, among other high-security settings, where the security and privacy of sensitive data are paramount.

Resource sensitivity, categorisation, and user security clearances are the primary factors in MAC access decisions. Based on the user's security level and the resource's sensitivity level, it grants or denies access in a hierarchical manner.

Key features of MAC include:

- **Label-Based Access Control:** To specify the degree of protection for both users and resources, MAC makes use of labels. Top Secret, Secret, and Unclassified are some examples of possible label classifications.

- **Security Clearances:** Based on their trustworthiness and the access they need, users are given security clearances.

- **Strict Enforcement:** MAC's stringent enforcement of access controls guarantees that only authorised users have access to sensitive resources.

Although MAC offers a great deal of protection, it can be difficult to set up and keep under control. A centralised authority is needed to establish and enforce access policies, which could reduce the system's agility and adaptability.

Systems managing sensitive client data or classified information systems are two examples of contexts where MAC is well-suited because of the extreme priority placed on data secrecy and integrity. It makes ensuring that critical resources may only be accessed by authorised individuals who have the proper security clearances.

The process of MAC implementation includes creating security classes, allocating security clearances, and setting up the system to enforce access policies according to these clearances and classifications.

Access Control Lists (ACLs)

The ACL is a crucial part of the ACL system, which is of access control mechanisms. These are utilised for the purpose of defining the level of access and who can access specific resources.

Permissions like read, write, and execute can be assigned to individuals or groups in an ACL list. The ACL is a database structure that defines the rights that users and groups have on various resources. Each item in the database represents a resource.

Key features of ACLs include:

- **Granular Access Control:** ACLs provide granular control over who can access what. Administrators can set permissions for certain users or groups with their help.

- **Dynamic Permissions:** Access permissions can be changed as needed thanks to the dynamic modification of ACLs. This adaptability is especially helpful in settings where access needs could fluctuate regularly.

- **Explicit Allow and Deny:** The ability to expressly grant or revoke access to resources is provided by ACLs. This feature enables administrators to set permissions for individual users or groups and limit their access to specific resources.

In systems that necessitate resource-level access control, such as file systems and network devices, ACLs are frequently employed. They offer a versatile and effective method for controlling who can access what resources by establishing and maintaining access permissions.

The steps to implement access control lists (ACLs) are to identify the resources that need protection, create entries in the ACL that detail the rights for each resource, and then assign users or groups to those entries.

To keep access permissions in line with the evolving needs of the organisation and to reduce the danger of illegal access, it is vital to evaluate and update ACLs on a regular basis.

Implementing Access Control Mechanisms

To guarantee a system's security, it is vital to implement access control mechanisms. Several processes are involved, including authorisation, authentication, and choosing the right access control models.

When users are authenticated, their identities are confirmed, and when they are authorised, their actions and resources are defined. Models for access control that offer frameworks for managing permissions include RBAC, DAC, MAC, and ACLs.

A few best practices that developers should adhere to when building access control mechanisms are conducting regular reviews and updates to access policies, utilising multi-factor authentication, and monitoring and auditing access control systems. These safeguards aid in preventing data breaches, unauthorised access, and noncompliance with security standards.

Authentication and Authorization

The implementation of access control mechanisms relies heavily on authentication and authorisation. They complement one another to restrict access to system resources to authorised users only.

Verifying a user's or system's identity is known as authentication. Passwords and other forms of authentication are required, and their validity is checked against a database of previously saved user details. Additional security measures may be implemented by using biometrics or security tokens as authentication elements.

The ability of an authenticated user to access certain resources and perform certain activities is determined by authorisation. Based on the user's identification and predefined rights or responsibilities, access permissions are granted or denied. Rules, regulations, or access control models (such as RBAC or DAC) may form the basis of authorisation.

The safety and soundness of any system depends on its authentication and authorisation procedures being well-implemented. Recommended methods consist of:

- Making use of strong, distinct passwords or introducing multi-factor authentication

- User responsibilities and permissions should be reviewed and updated on a regular basis.

- Establishing safe procedures for the transmission of authentication data

- Giving users only the rights, they need, in accordance with the concept of least privilege

Developers may lessen the likelihood of data breaches and unauthorised access by establishing strong authentication and authorisation procedures to restrict access to sensitive resources to only authorised users.

Access Control Models

Manage and enforce access rights with the use of Access Control Models, which are frameworks used in the Implementation of Access Control Mechanisms. These models provide forth a methodical framework for establishing the parameters and requirements for resource access.

Some commonly used access control models include:

- **Role-Based Access Control (RBAC):** RBAC uses established roles to assign access rights. Permissions are provided to users depending on the roles they are allocated, which are based on their job duties.

- **Discretionary Access Control (DAC):** Resource owners may manage their resources' access permissions using DAC. Owners have the authority to allow or prohibit access to certain people or groups at their own discretion.

- **Mandatory Access Control (MAC):** MAC uses labels and specified rules to enforce access controls. The resource's sensitivity and the user's security clearance are the two main factors considered when deciding who has access.

Other access control models include ABAC and RBAC.

One must consider the system's characteristics and unique security needs while deciding on an access control approach. When making your decision, keep in mind the following: scalability, flexibility, management convenience, and required control level.

Developers may lessen the likelihood of illegal access and keep the system secure and intact by using the correct access control model to monitor and enforce access rights.

6.3 Compliance with Data Privacy Regulations

The goal of database security is to prevent data loss, corruption, or alteration by implementing appropriate rules and procedures. The purpose of database security rules is to protect private information from unauthorised access and to guarantee the consistency and availability of database records. Because of the high stakes involved in data breaches and illegal data manipulation, many businesses are making database and storage security a top priority.

Database Security in Public Clouds

Database services, both managed and unmanaged, are available from public cloud providers. Customers with unmanaged or semi-managed databases would be responsible for maintaining them on their own virtual machines, while cloud providers with managed databases implement security updates, update software, and ensure high availability. (The cloud service would continue to handle management of the underlying physical infrastructure.)

The shared responsibility concept is followed by cloud database security in both instances. When you use managed database services, your cloud provider will also take care of patching, upgrading, and monitoring for any possible security vulnerabilities with the underlying infrastructure, which includes compute, storage, and networking resources.

Organisations, on the other hand, need to comply with applicable legislation, construct appropriate access controls, and secure the data kept in databases. To that end, it is necessary to encrypt critical data, set up permissions for database access, keep an eye out for suspicious activity, and educate staff on best practices for security.

Agentless security technologies are often necessary for organisations due to the absence of access to cloud database servers. Rather of installing software on the actual database, these technologies work remotely and monitor the database using APIs and data extracts. One further benefit of agentless security technologies is that they won't slow down the cloud or use up too many resources. Databases may be adequately protected without adding extra resource overheads with the help of agentless security solutions that provide real-time threat detection, vulnerability scanning, and compliance management.

Elements of Database Security

- **Authentication and Identity Management**

  Applying robust authentication methods such as multifactor authentication (MFA), single sign-on (SSO), and appropriate role-based access control (RBAC) to guarantee that the database can only be accessed by authorised users. Users' permissions and the data they may access inside the database can be fine-tuned in this way.

- **Data Encryption**

  Data encryption using established techniques ensures protection both while data is at rest and when it is in transit. This aids in making sure that data stays unintelligible without the proper decryption keys, even if unauthorised access happens.

- **Data Masking and Redaction**

  Data masking and redaction refer to the practice of concealing sensitive information from users without appropriate authorisation. To prevent unauthorised parties, even those with restricted access, from gaining access to sensitive information, data might be anonymised, pseudonymized, or obfuscated.

Intrusion Detection and Prevention

The database is being monitored and protected from malicious activity by putting intrusion prevention systems (IPS) and intrusion detection systems (IDS) into place. Cross-site scripting, SQL injection, and other efforts to take advantage of flaws in the database system may be detected and stopped with the use of these technologies.

Backup and disaster recovery: The best way to protect data from corruption or loss caused by hardware failure, software mistakes, or malicious activity is to back it up regularly and have a solid disaster recovery plan. A safe and working database may be swiftly restored in the event of an incident thanks to this.

- **Patch Management**

  The best way to safeguard your database from security holes and known vulnerabilities is to always use the most recent fixes and upgrades. Reducing the possibility of exploitation is possible via the implementation of a patch management strategy, which guarantees the consistent and timely application of updates.

- **Access Control**

  Using the concept of least privilege and other granular access control rules to limit user rights to those that are absolutely essential for their job. By regulating the privileges of each user in the database, this reduces the possibility of data loss or alteration due to unauthorised access. (See: What is data access governance?)

- **Network Security**

  Taking precautions to protect the network infrastructure that links to the database, such as installing firewalls and VPNs. This helps keep the database and the rest of the company's systems from being accessed by unauthorised parties by protecting the routes of communication between the two.

- **Monitoring and Alerting**

  Consistently keeping an eye on database health, user actions, and any security risks to spot irregularities or breaches. By using real-time alerts, the security team may be promptly notified of any suspicious actions, allowing them to respond quickly and minimise risks.

Database Security: 8 Best Practices

In order to prevent data theft and unauthorised access, organisations should follow certain database security best practices. Here are a few of these recommended methods:

- Database security rules and procedures should be reviewed and updated on a regular basis.

- Identify and resolve any security weaknesses by monitoring database operations and generating security reports.

- To stop unauthorised people from getting in, use MFA, RBAC, and strong passwords.

- Data should be encrypted both while it is in transit and while it is stored to prevent unlawful access or alteration.

- To make sure database software is safe and up-to-date, it's important to apply patches and upgrades immediately.

- Regular backups and a plan for when calamity strikes should be your top priorities.

- It is important to find any possible weak spots in the database security posture by doing vulnerability scans and risk assessments.

- Train employees to understand their part in keeping databases secure and why it's critical.

Multiple Choice Questions (MCQs)

1. **Which of the following encryption techniques is commonly used to protect data at rest?**

    a. AES (Advanced Encryption Standard)

    b. RSA

    c. SHA (Secure Hash Algorithm)

    d. Diffie-Hellman

2. **Which access control model is based on roles assigned to users within an organization?**

   a. Discretionary Access Control (DAC)

   b. Role-Based Access Control (RBAC)

   c. Mandatory Access Control (MAC)

   d. Attribute-Based Access Control (ABAC)

3. **Which of the following data privacy regulations focuses on protecting the privacy of individuals within the European Union?**

   a. HIPAA (Health Insurance Portability and Accountability Act)

   b. CCPA (California Consumer Privacy Act)

   c. GDPR (General Data Protection Regulation)

   d. PCI DSS (Payment Card Industry Data Security Standard)

4. **Which of the following is a key benefit of using public key encryption (asymmetric encryption)?**

   a. It requires less computational power than symmetric encryption

   b. It uses a single key for both encryption and decryption

   c. It enables secure communication over an insecure channel

   d. It is not suitable for large datasets

5. **Which of the following is an example of an identity-based access control mechanism?**

   a. Two-factor authentication

   b. Biometric verification

   c. IP address filtering

   d. Both A and B

6. **Under GDPR, which of the following rights allows individuals to request the deletion of their personal data?**

   a. Right to Access

   b. Right to Rectification

   c. Right to Erasure (Right to be Forgotten)

   d. Right to Data Portability

7. **In which scenario would the use of end-to-end encryption be most appropriate?**

   a. Sending a plain text email without any sensitive information

   b. Transmitting sensitive financial data over an untrusted network

   c. Storing user passwords in a local database

   d. Creating an access control list for a file

8. **Which of the following describes "least privilege" in access control?**

   a. Users are given the least amount of data to perform their tasks

   b. Users can only access data necessary for their roles

   c. Users are given complete access to all systems

   d. All users have equal access to the system

9. **Which of the following is a requirement under the CCPA for businesses collecting personal information in California?**

   a. Businesses must provide consumers with the option to opt out of the sale of their data

   b. Businesses must store all personal data for at least 10 years

   c. Businesses are allowed to collect personal data without any restrictions

   d. Consumers must be automatically included in marketing emails

10. **What is the purpose of using a cryptographic hash function in data security?**

    a. To encrypt data and make it unreadable without a key

    b. To generate a fixed-size output from variable-length input, ensuring data integrity

    c. To encrypt messages using a public key

    d. To perform symmetric encryption with the same key for both encryption and decryption

Answer

1	2	3	4	5	6	7	8	9	10
A	B	C	C	D	C	B	B	A	B

ADVANCED TOPICS IN DATA ENGINEERING

7.1 Machine Learning Integration

Machine learning has become an essential part of data engineering, helping to uncover deep insights in large and complicated datasets. The use of machine learning methods has become crucial in understanding patterns, producing forecasts, and allowing informed decision-making due to the exponential growth of data volume. The convergence of data engineering and machine learning improves data processing and paves the way for novel solutions in many different sectors. This section explores the core elements of the relationship between ML and data engineering, illuminating its importance and diverse uses.

Fundamentals of Machine Learning

A solid grasp of fundamental ideas is necessary when working with Machine Learning (ML) in the context of Data Engineering. The fundamental idea behind ML is to use automated methods to derive insights from data. Underlying this area are numerous essential concepts:

Three main learning paradigms are used in machine learning: supervised, unsupervised, and reinforcement learning. Predictions and classifications are made possible in supervised learning systems by training them on labelled data. Contrarily, unsupervised learning seeks to uncover patterns or structures within data without labels. By using a series of trial-and-error judgements to optimise for a target, algorithms may be trained to perform reinforcement learning.

The separation of data into two groups, one for training and one for testing, is an essential part of machine learning. Algorithms learn patterns and correlations from training data, which helps with their generalisation, and they are evaluated on testing data, which helps with their performance on new, unknown data.

An essential part of ML is algorithm selection. From SVMs and clustering approaches to decision trees and neural networks, each algorithm serves a unique function. Both the issue and the dataset properties should be considered while deciding on an approach.

In the end, understanding these basic ideas lays a strong foundation for understanding how machine learning fits into the bigger picture of data engineering. With this information in hand, experts can create and implement data pipelines that efficiently use machine learning to glean insights from complicated datasets.

Integration of Machine Learning and Data Engineering

When it comes to data engineering and machine learning, two crucial steps are merging together: preparing data and building efficient pipelines.

- **Data Pre-processing for Machine Learning**

  Data has to be pre-processed thoroughly before machine learning algorithms can be used. Data cleansing, the process of correcting mistakes and inconsistencies, and data transformation, the process of transforming data into an analytically viable format, are both part of this process. To further enhance the model's performance, feature engineering and selection are carried out to determine which features are most important. The results of the following machine learning procedures may be improved in terms of accuracy and significance by properly preparing the data.

- **Data Pipelines for Machine Learning Workflows**

  Machine learning processes rely on data pipelines, which coordinate the transformation and transfer of raw data to an analytically suitable state. In a typical pipeline, steps like data extraction, data transformation, and data loading bring together data from many sources, reshape it, and then put it into the target environment for analysis. Pipelines facilitate the effective construction and deployment of machine learning models by ensuring that data flows smoothly through the pre-processing, training, and evaluation phases.

  There is a mutually beneficial link between data engineering and ML when both components are integrated effectively. This allows data engineers to use machine learning to extract insights and make predictions from complicated datasets.

Challenges and Considerations

- **Scalability and Performance Issues**

  Scalability and performance issues may arise when data engineering methods include machine learning. Efficiently processing and analysing large datasets becomes more challenging as their sizes increase. Making sure that machine learning algorithms can process massive amounts of data with tolerable latency is of the utmost importance. To overcome these obstacles, data engineers must optimise the foundational infrastructure, including distributed computing frameworks. To maintain system efficiency as data grows, scalability solutions are crucial, such as clustering and parallel processing.

- **Handling High-Dimensional Data**

  The large dimensionality of many real-world datasets presents difficulties for machine learning techniques. The "curse of dimensionality" causes problems with overfitting, higher computing needs, and identifying significant patterns. Data scientists and data engineers need to collaborate to find solutions to these problems by using dimensionality reduction methods like feature selection or Principal Component Analysis (PCA). An important factor to think about is the methods to use to decrease dimensions while keeping important data.

- **Model Drift and Retraining Strategies**

  It is possible that future data patterns may differ from the ones used to train ML algorithms. When data distributions change or when other external variables cause the model's performance to decline over time, this is called model drift. Data engineers must devise methods to track how well models are doing, identify instances of drift, and retrain them as needed. To keep their accuracy and relevance, models may adapt to changing situations via continuous data gathering and integration into training.

Tools and Technologies

- **Frameworks for Implementing Machine Learning Pipelines**

  Data preparation, model training, and deployment are three interconnected but distinct processes that must be managed by ML pipelines. Data engineering procedures may be easily integrated with machine learning using many common frameworks.

  - **TensorFlow:** A machine learning library created by Google, TensorFlow is available to the public as open-source software. It provides a versatile environment for developing and deploying a wide range of ML models, including neural networks with little to no complexity. In many contexts, its adaptability makes it the material of choice.

  - **PyTorch:** PyTorch is a popular open-source framework for deep learning. Due to its user-friendly layout and compatibility with dynamic neural networks, as well as its dynamic computation graph, it is highly regarded by academics and developers.

- **Scikit-learn:** An efficient and straightforward machine learning library, Scikit-learn prioritises user experience. For smaller-scale projects or rapid prototypes, it offers a variety of tools for data preparation, feature selection, and model validation.

- **Big Data Technologies for Processing and Storing ML Data**

  Specialised Big Data solutions are needed to manage massive amounts of data from machine learning as the volume of data keeps on increasing. To integrate machine learning effectively, these technologies guarantee efficient processing and storage of big datasets.

  - **Apache Spark:** Distributed data processing has never been easier than using Apache Spark, an open-source platform. It offers libraries for a wide range of uses, such as processing graphs, ML, and data preparation. It can speed up calculations by a large margin due to its in-memory processing capabilities.

  - **Hadoop:** Hadoop is a popular Big Data system that incorporates the MapReduce programming approach for handling massive datasets and the HDFS for distributed storage. Although it has a large user base, Spark and other newer frameworks have recently surpassed it in popularity because to their superior performance.

  - **Distributed Databases:** Distributed databases are specifically engineered to handle massive amounts of data over several nodes. Some examples of such databases include Google Bigtable, Apache Cassandra, etc. These database systems provide the availability, scalability, and fault tolerance of data, which are prerequisites for machine learning jobs.

- **Collaboration between Data Engineers and Data Scientists**

  Data scientists and data engineers work hand in hand to maximise data engineering's potential for machine learning in the ever-changing world of data-driven solutions. The development of effective machine learning applications depends on bridging the gap between these two critical functions.

- **Bridging the Gap between Data Engineering and Data Science Roles**

  Data scientists and data engineers need to coordinate their efforts to maximise the potential of machine learning. Data engineers provide the groundwork for processing, storing, and collecting data. Data scientists are busy creating and implementing machine learning models to get insights from the data. Working together, we can guarantee that the models will be more accurate and efficient since they will be based on well-optimized data pipelines.

- **Effective Communication and Collaboration**

  An effective partnership between data scientists and data engineers relies on open lines of communication. Each side has a better grasp of the other's data needs, model limitations, and business goals via frequent communication. Experts in data science may provide light on the ML model's complexities, input expectations, and intended results, while data engineers can offer insights on the data's availability, quality, and transformations required for analysis. Teams are better able to tackle problems and refine their solutions when they share knowledge openly.

  A company may get the most out of its data engineers and data scientists by encouraging a cooperative work environment. The development of data-driven apps that benefit both companies and consumers is made possible by this partnership, which guarantees robust and precise machine learning models that are also easily incorporated into the data engineering framework.

Future Trends and Developments

- The development of more effective and precise machine learning algorithms.

- The automation of machine learning processes from beginning to finish has been enhanced.

- Consolidating data pipelines with AI-driven decision-making for insights in real-time.

- A rising awareness of the need of ethical issues in data engineering and machine learning.

- Machine learning activities will be accelerated via the continued development of specialised hardware.

- Machine learning models are now easier to understand and explain.

- Research on methods for protecting sensitive information via federated learning and similar initiatives.

- The merging of big data with machine learning to provide increasingly more complex studies.

Data engineering is greatly enhanced by machine learning, which allows for the extraction of meaningful insights from massive datasets. To remain at the forefront of using the synergy between data engineering and ML, it is crucial to continuously learn and adapt, since this subject is always changing. A proactive strategy for talent development and innovation is crucial as these fields keep influencing data-driven solution futures.

7.2 Big Data Technologies

Big Data Technology is the key that unlocks the door for tech behemoths like Amazon and Apple to become an integral part of our daily life. Operational analytics, sales management, and supply chain efficiency all benefit from this technology. There are essentially two technologies that may be employed with big data, and those technologies can be further separated into four key areas.

The term "Big Data Technology" describes a group of programs that help handle various datasets and make sense of them for commercial purposes. This system processes, analyses, and extracts useful information from massive datasets with complicated architectures. There is a strong relationship between big data and cutting-edge technologies such as the IoT, ML, and AI. (Simplilearn, 2024)

Applications of Big Data Technologies

There is a wide variety of industries that may benefit from big data technologies. The following are examples of known fields of application:

- **Healthcare:** Patients' data is analysed using Big Data technology in order to create individualised treatment programs. In addition to optimising healthcare operations, it provides predictive analysis for illness outbreaks and effectively devises treatment programs.

- **Finance:** Insights into the financial sector for fraud detection are greatly enhanced by this technology. In addition, it helps break down the target market into several types of customers.

- **E-Commerce:** The advent of useful recommendation engines made possible by big data technology has revolutionised the way consumers purchase.

- **Education:** This technology provides insights into analytics of student performance and aids in the creation of adaptable learning platforms for individualised education.

- **Retail:** Retailers may analyse consumer behaviour with the use of Big Data Technology to create targeted marketing campaigns. Managing inventories and optimising prices according to market trends are other key areas of concentration.

Types of Big Data Technologies

There are a number of distinct kinds of big data technologies, which, when combined, have many potential applications across the data lifecycle.

- **Data Storage:** One crucial part of data processing is data storage. Gathering, storing, and organising massive amounts of data for easy access is the main responsibility of some big data technologies.

- **Data Mining:** Data mining is a powerful tool for uncovering previously unseen trends and patterns, which in turn lead to deeper knowledge. In order to extract valuable information from raw data sets, data mining technologies use a variety of statistical methodologies and algorithms. Apache Storm, Flink, ElasticSearch, Rapidminer, and Presto are some of the best big data solutions for data mining operations.

- **Data Analytics:** Utilising artificial intelligence, big data solutions with powerful analytical skills may provide valuable insights for businesses and supply the data needed to make important choices.

- **Data Visualization:** It is crucial to simplify and make useful this information since big data technologies deal with large amounts of data that is organised, semi-structured, unstructured, and has varying degrees of complexity. Data presented visually, in the form of graphs, charts, or dashboards, is more interesting and simpler to understand.

Figure 7.1: Big Data Technologies

Source: - *(Vyas, 2023)*

Top Big Data Technologies

a. Apache Hadoop

Distributed storage and processing of large data sets using simple programming methods are made possible by the open-source Apache Hadoop framework. The HDFS and the MapReduce programming paradigm are components of it. Due to its design, Hadoop can expand from a few servers to thousands of devices, all of which can perform computation and storage on their own. Hadoop is an important tool for managing large-scale data processing jobs because it effectively handles massive volumes of structured and unstructured data. It is a foundational technology in the big data environment.

b. Apache Spark

Apache Spark is a free and open-source unified analytics engine that processes large amounts of data quickly and easily. In contrast to Hadoop MapReduce, which is based on discs, it offers in-memory computing capabilities, which greatly improve the speed of huge data processing operations. Spark provides high-level APIs for tasks including SQL queries, streaming data, ML, and graph processing; it is compatible with Scala, Java, Python, R, and similar languages. An adaptable component of the big data ecosystem, it may handle data in batches or in real time.

c. Apache Kafka

When it comes to real-time data streams, Apache Kafka is the platform to manage them. With its focus on low-latency, high-throughput data processing, Kafka was first developed by LinkedIn. Data producers may use it to transmit records to Kafka topics, and consumers can read them. This paradigm is useful for creating streaming applications and real-time data pipelines. With its scalable architecture, Kafka is perfect for real-time data processing applications including stream processing, real-time analytics, and log aggregation. It can process millions of messages per second.

d. Apache Flink

One open-source framework that can process data in both real-time and batch formats is Apache Flink. Throughput and latency are both much improved, and it can accurately compute stateful operations across both finite and unbounded data streams. Complex event processing, graph processing, and machine learning are some of Flink's high-end features. Applications requiring processing of massive amounts of data will find it to be an ideal fit because to its scalable and fault-tolerant design. Applications that need to analyse continuous data flows may find Flink's powerful windowing and state management features especially handy.

e. Google Big Query

A serverless data warehouse that is under complete management and uses Google's infrastructure to speed up SQL queries. With it, you can efficiently query massive datasets without having to worry about managing the underlying infrastructure. For optimal speed and scalability, Big Query uses a distributed architecture in conjunction with a columnar storage format. Because of its real-time data analysis capabilities and ability to interface with other Google Cloud services, it is a crucial tool for machine learning, data analytics, and business intelligence applications.

f. Amazon Redshift

Easy SQL and BI tool analysis of massive datasets with this completely managed cloud data warehouse solution. You can execute sophisticated analytical queries on petabytes of structured and semi-structured data with Redshift since its architecture is built for high-performance queries. It has capabilities that improve performance, like as compression of data, columnar storage, and parallel execution of queries. Because of its compatibility with a wide range of analytics tools and data sources, Redshift is an adaptable option for BI and big data projects.

g. Snowflake

Scalability, speed, and user-friendliness are the hallmarks of the Snowflake cloud data warehousing platform. In order to provide independent scalability and maximised performance, Snowflake's design distinguishes between storage and computing resources, unlike conventional data warehouses. With its powerful SQL capabilities, it allows for the querying and analysis of both structured and semi-structured data. Organisations of any size may benefit from Snowflake's multi-cluster design, which guarantees high concurrency and effective workload control. Its adaptability in the big data ecosystem is boosted by its smooth connection with a variety of cloud services and data integration tools.

h. Databricks

Databricks is an Apache Spark-based analytics platform that brings together data science, engineering, and business to speed up innovation. Data teams may work together in this collaborative environment to analyse and learn from massive amounts of data. The data pipeline development and deployment process is made easier with Databricks' optimised runtime for Apache Spark, interactive notebooks, and integrated data processes. Big data analytics and AI-driven applications may use its tremendous capabilities to manage batch and real-time data.

i. MongoDB

The NoSQL database MongoDB is well-known for its user-friendliness, scalability, and adaptability. A more natural and adaptable data format than conventional relational databases is made possible by storing data in documents similar to JSON. Whether you're using it for content management, the IoT, or real-time analytics, MongoDB can handle big amounts of data, both organised and unstructured. Supporting complex data relationships and great performance, it has the potential to scale horizontally and a comprehensive query language.

j. Cassandra

Distributed and very scalable, Apache Cassandra is a NoSQL database designed to handle massive amounts of data across several commodity servers in a failsafe fashion. Perfect for mission-critical apps, its decentralised design ensures high availability and fault tolerance. You can easily manage varied data types with Cassandra's support for flexible schemas and its ability to handle structured and semi-structured data. Its continuous performance, guaranteed by linear scalability, makes it well-suited for use cases including online transaction processing, real-time analytics, and the Internet of Things.

Emerging Big Data Technologies

A number of other big data technologies are on the horizon, in addition to the ones already listed. Some of the most important technologies in this category are as follows:

- **TensorFlow:** TensorFlow is a community resource that academics may use to create state-of-the-art Machine Learning. It mixes many comprehensive libraries with flexible ecosystem tools. Above importantly, this paves the way for developers to create and launch environments-specific apps backed by machine learning.

  The Google Brain Team launched TensorFlow in 2019. Python, CUDA, and C++ form its basic foundation. Google, eBay, Intel, and Airbnb are among the companies using this technology to meet their commercial needs.

- **Beam:** Apache Beam is an API layer that can be easily transported and used to construct and manage complex pipelines for processing data in parallel. On top of that, it enables the use of various execution engines or runners to execute created pipelines.

  The Apache Software Foundation released Apache Beam in June 2016. Python and Java made it possible. This innovation is being used by certain major corporations, such as Verizon Wireless, Amazon, ORACLE, and Cisco.

- **Docker:** Docker is a software platform that facilitates application development, deployment, and execution via the use of containers. Commonly, containers aid programmers in correctly packaging apps with all of their necessary components, such as libraries and dependencies. Usually, components are bound together and sent in a single box by means of containers.

  In March 2003, Docker Inc. released Docker. The Go language serves as its foundation. Splunk, Business Insider, Quora, and PayPal are among the companies that use this technology.

- **Airflow:** As a scheduling and workflow automation solution, Airflow is a technological marvel. The primary function of this technology is to manage and keep data pipelines running smoothly. It includes various tasks organised into workflows that make use of the Directed Acyclic Graphs (DAGs) concept. To facilitate testing, maintenance, and versioning, developers may also specify processes in codes.

  The Apache Software Foundation released Airflow in May of 2019. Python is its foundational language. This cutting-edge technology is being used by Checkr and Airbnb, among others.

- **Kubernetes:** Google released their open-source Kubernetes platform for managing clusters and containers in 2014. It lays the groundwork for host cluster application container operations, deployment, scalability, and automation.

The Kubernetes platform was launched by the Cloud Native Computing Foundation in July 2015. You may read it written in Go. North-western Mutual, Pear Deck, People Source, and American Express are among the companies that are benefiting from this technology.

These are new forms of technology. The big data ecosystem is always evolving, however, so they aren't constrained. Because of this, the demand and needs of the IT sectors are driving the rapid development of new technologies.

7.3 Data Engineering in the Cloud

An expert in information technology, a Cloud Data Engineer plans, constructs, and oversees the systems and software required to process and analyse massive amounts of data on the cloud. Their knowledge of databases, data warehousing, and ETL (extract, transform, load) procedures is extensive, and they are specialists in cloud computing platforms like AWS, Azure, or Google Cloud. By making sure data is easily available, trustworthy, and optimised for performance, these engineers are crucial in allowing data-driven decision-making. Cloud Data Engineers are the brains behind the data pipeline, making sure all the data goes from different sources to the people who need it. This helps businesses get insights from big data and stay ahead of the competition online.

What does a Cloud Data Engineer do?

Data processing and storage solutions on the cloud are the speciality of cloud data engineers, who also design, create, and manage them. To ensure data is accessible, safe, and processed rapidly for analysis and actionable insights, they play a key role. Organisations may optimise data pipelines, expand data infrastructure, and enhance data-driven decision-making with the use of cloud technology.

Key Responsibilities of a Cloud Data Engineer

- Creating and executing cloud-based data storage solutions that are both safe and scalable, with an eye on maximising performance and user accessibility.

- The creation and upkeep of reliable data pipelines for the processing, transformation, and dissemination of massive datasets.

- Automation of data workflows and optimisation of data engineering processes via the use of cloud services and related technologies.

- Assisting with data modelling, analysis, and reporting requirements via collaboration with data analysts, data scientists, and other stakeholders.

- Making sure that data encryption and access restrictions are in place and that all data governance and security standards are followed.

- Tracking the efficiency of cloud data systems, finding wasteful processes, and fixing them.

- Taking steps to guarantee the correctness and integrity of data and performing data quality checks.

- Building application programming interfaces (APIs) for data consumption by different applications or users and optimising data retrieval.

- Keeping abreast of new developments in data engineering and cloud computing so that we may suggest and implement improvements to our data systems.

- Offering assistance and guidance pertaining to technical matters with data, such as identifying and fixing problems with data pipelines.

- Disaster recovery plans for data stored in the cloud should be developed and implemented in conjunction with the relevant IT and security departments.

- Data flow diagrams, process documentation, and metadata maintenance for data lineage and cataloguing are all part of data engineering.

Day to Day Activities for Cloud Data Engineer at Different Levels

A Cloud Data Engineer's day-to-day duty could vary greatly according on their degree of expertise. While entry-level engineers tend to concentrate on data infrastructure support and technical skills, middle-level engineers take on more complicated projects and begin to specialise. Cloud Data Engineers at the senior level are required to spearhead projects, create data architecture, and pitch in with strategic choices. In guiding the company's data strategy, they are essential.

- **Daily Responsibilities for Entry-Level Cloud Data Engineers**

  Cloud data systems rely on the development and upkeep of entry-level Cloud Data Engineers' technical abilities. Their day-to-day tasks often need tight supervision and teamwork with more seasoned colleagues.

  - Monitoring and troubleshooting data pipelines and workflows

  - Assisting in the implementation of data storage solutions

  - Performing basic data transformations and batch processing

  - Learning cloud service providers' tools and technologies (e.g., AWS, Azure, Google Cloud)

  - Documenting data processes and maintaining data dictionaries

  - Participating in code reviews and learning best practices

- **Daily Responsibilities for Mid-Level Cloud Data Engineers**

  Mid-level Cloud Data Engineers are responsible for a greater range of tasks, including the management of bigger data infrastructure segments and the optimisation of data processing, storage, and retrieval systems. They take greater initiative and may start to focus on a certain field of technology or business.

  - Designing and constructing robust data pipelines

  - Implementing data security and compliance measures

  - Maximising efficiency and minimising expenses in data storage and processing

  - Developing custom data models and algorithms

  - Working together with analysts and data scientists to back analytics projects

- The integration of third-party services and the automation of data processes

- **Daily Responsibilities for Senior Cloud Data Engineers**

  Experts in their profession, Senior Cloud Data Engineers steer data engineering initiatives strategically. To make sure data strategies are in line with company objectives, they supervise complicated projects, guide younger engineers, and collaborate closely with other senior technical personnel and management.

  - Crafting data solutions that are safe, scalable, and in line with company goals

  - Coordinating the efforts of interdisciplinary teams to complete massive data projects

  - Analysing data at a high level and providing strategic suggestions

  - Documenting and implementing data governance and lifecycle management best practices

  - Promoting creativity and investigating novel data tools and approaches

  - The data engineering team's mentorship and development

Key Cloud Data Engineering Tools and Features

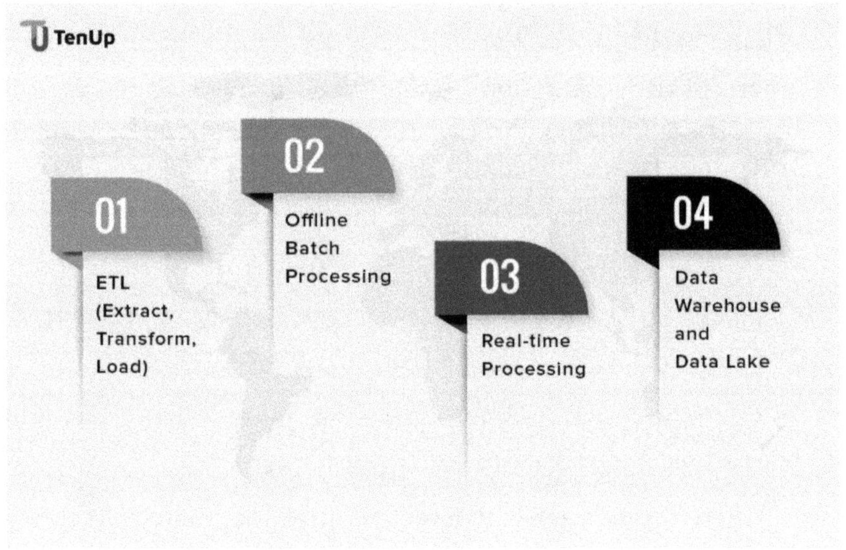

Figure 7.2: Cloud Data Engineering Tools and Features

Source: - *(Parikh, 2023)*

1. ETL (Extract, Transform, Load)

The ability to extract, convert, and load data from several sources into a destination—like a data warehouse or a data lake—is why ETL tools are crucial in cloud data engineering. By using these technologies, data engineers may effectively gather information from various sources like as databases, applications, and APIs. They can then apply the appropriate transformations to the data and store it in the right location for further analysis and reporting.

It is possible for data engineers to convert data from one system into another's format with the use of ETL technologies, which provide powerful data mapping and transformation capabilities. When it comes to ETL, these solutions have you covered with a variety of transformation operations that cleanse and enhance your data, including filtering, aggregating, joining, and more.

- **ETL Scheduling and Orchestration:** Data engineers may automate the construction of data pipelines and set them to run at predetermined intervals or in reaction to events using the scheduling and orchestration tools included in ETL systems. Automated data retrieval, translation, and input into the target system is now possible thanks to this feature, which greatly improves efficiency by eliminating the need for human intervention.

2. Offline Batch Processing

Cloud data engineering relies on Offline Batch Processing, which involves processing large data sets at regular intervals. Data engineers may better handle massive datasets with the help of cloud-based batch-processing tools, which partition the information into smaller, more manageable pieces and analyse them in parallel. This strategy is great for situations when data latency is tolerable and real-time processing is not necessary since it maximises resource utilisation and cost-effectiveness.

- **Job Recovery and Fault Tolerance:** Data engineers can recover from processing failures and mistakes thanks to built-in fault tolerance features in cloud batch processing solutions.

- **Scalability and Parallel Processing:** With this scaling capability, batch tasks may handle ever-increasing data volumes without sacrificing speed, allowing them to analyse big datasets in manageable periods. These tools' parallel processing capability guarantees seamless data processing regardless of data size growth and maximises efficiency.

3. Real-time Processing

Data engineers may react in real-time to incoming data streams with the help of real-time processing tools, which are an integral aspect of business cloud solutions. This allows for lightning-fast insights and decision-making. Rapid reactions based on fresh data are needed in many applications, such as real-time analytics, fraud detection, and

IoT data processing, among others. This capability is crucial for these environments.

- **Low Latency and Event Time Processing:** Quick responses to data insights and events are made possible by real-time processing technologies, which are very effective in low-latency data analysis. For real-time applications, its support for event time processing guarantees that data is processed in the order in which they happened, allowing for precise analysis of time-sensitive data while maintaining data integrity.

- **Windowing and Time-Based Aggregation:** Continuous data analysis and presentation are made possible by this feature, which allows for the building of time-based aggregates like hourly, daily, or sliding window aggregates.

4. Data Warehouse and Data Lake

Since data warehouses are designed to optimise query efficiency and structured data processing, they are well-suited for reporting and business intelligence. In contrast, data lakes are designed to store vast quantities of unstructured and raw data, as well as other types of data, and to facilitate sophisticated analytics, ML, and data exploration.

- **Schema-on-Read and Schema-on-Write:** The schema-on-read technique is useful for data lakes because it allows for the storage of data in its raw, unstructured state. Data scientists and data engineers may use the schema for data analysis due to its versatility, which makes it suitable for dealing with semi-structured and unstructured data. In contrast, data warehouses often use the schema-on-write approach, which necessitates pre-intake data formatting for best query speed.

- **Query Optimization and Materialized Views:** Data warehouses often use query optimisation methods and materialised views to enhance the performance of queries. Caching the results of frequently executed queries in materialised views improves system performance and decreases query response times, especially for complex analytical queries on big datasets.

Types of Cloud Data Engineers

An essential part of managing, processing, and analysing data in the cloud is cloud data engineering, a dynamic and quickly developing subject. There are now many subspecialties within the field of Cloud Data Engineering due to the high need for specialised skillsets caused by the widespread use of cloud computing by businesses. Different types of Cloud Data Engineers are responsible for various parts of cloud data ecosystems and bring unique skill sets to the table. Data accessibility, security, and optimum utilisation to generate business insights and choices are critically dependent on their knowledge. Numerous career paths are possible due to the variety of positions, which meets the unique requirements of various cloud data projects and organisational objectives.

- **Infrastructure Cloud Data Engineer**

  Cloud data infrastructure is the domain of Infrastructure Cloud Data Engineers, who are experts in its design and management. They are well-versed in all things related to cloud computing, from storage solutions and data lakes to distributed computing resources and cloud architecture. Making sure data platforms are dependable, fast, and scalable is their main concern. They collaborate closely with cloud service providers and IT departments to set up and keep running the systems needed to do analytics and data processing on a massive scale. For businesses, especially those facing big data issues, this position is essential in laying a solid groundwork for data operations.

- **Data Integration Engineer**

  Data Integration Engineers are masters in bringing together disparate data sets and storing them in a single, accessible location on the cloud. To guarantee smooth data flow and integration, they are experts in dealing with APIs, data pipelines, and ETL (Extract,

Transform, Load) technologies. Data governance and quality control, in addition to the technical parts of data input, are within their purview. When data has to be standardised and prepared for analysis from a variety of sources, they play a crucial role.

- **Cloud Data Warehouse Engineer**

  Data warehousing solutions hosted in the cloud are the speciality of cloud data warehouse engineers. They are well-versed in the features and capabilities of cloud data warehouse services and have extensive experience with data modelling, warehouse design, and SQL. Business intelligence and data-driven decision-making are made possible by optimising data storage for quick querying and reporting. They play an essential role in companies that get strategic insights from reporting and analysis of historical data.

- **Big Data Cloud Engineer**

  Massive Data Repository Cloud Engineers focus on managing massive data sets in the cloud. Hadoop, Spark, and NoSQL databases are all part of big data technology, and they are well-versed in using the scalability of the cloud to handle massive amounts of data. Companies rely on this knowledge when they need to derive insights from complicated information, often in real-time. Their focus is on creating and refining analytics apps and data processing tasks that can deal with the diversity, speed, and volume of big data.

- **Cloud Data Security Engineer**

  Data protection in the cloud is the primary focus of cloud data security engineers. Security, encryption, and access control are areas in which they excel. Installing security measures, keeping an eye out for breaches, and making sure everything is in line with data protection standards are all part of their job description. Particularly for sectors like healthcare and finance that deal with personal information, these cloud engineers work in tandem with other members of IT security teams to provide a safe data environment.

- **Machine Learning Data Engineer**

  Professionals specialising in machine learning data engineering help to connect the dots between the two fields. Datasets for training machine learning models are prepared and managed by them, and then these models are integrated into production settings. They need to be familiar with model deployment strategies, data pre-treatment methods, and machine learning frameworks to do their job well. An invaluable resource for businesses seeking to tap into the predictive potential of their data, they collaborate closely with data scientists and machine learning developers to bring AI-driven applications to life.

Benefits of Cloud Data Engineering Tools for Scalable Data Processing

- **Flexibility & Scalability:**

  Cloud data engineering tools enable businesses to adjust resource allocation in response to fluctuating demand via elastic resource provisioning and auto-scaling capabilities. This safeguards the efficient handling of processes that are heavy on data and do not care about hardware limitations.

- **Cost Effectiveness:**

  Data engineering solutions in the cloud work on a pay-as-you-go model, so you only pay for the resources you really use. To better allocate resources, this inexpensive approach does away with the need to invest in infrastructure up front.

- **Data Integration and ETL procedures:**

  With data engineering technology, ETL processes are simplified by integrating data from several sources without any noticeable hiccups. Data may be quickly changed and supplied into destination systems using effective ETL pipelines.

- **Cloud data engineering solutions:**

  allow for the storing and retrieval of data at the same time by using distributed data processing frameworks. Its enhanced processing speed and performance make it well-suited for processing activities involving massive amounts of data.

- **Real-time Data Processing:**

  Cloud data engineering tools provide stream processing, which enables low-latency data input and processing. With the help of real-time data insights, companies can quickly adapt to new data situations, which speeds up crucial decision-making.

What's it like to be a Cloud Data Engineer?

When you become a Cloud Data Engineer, you join an industry where data powers contemporary companies. In this field, analytical acuity and technological know-how come together to construct, manage, and enhance cloud-based data infrastructures that are robust, scalable, and safe.

Ensuring data quality, building and maintaining reliable data pipelines, and making data accessible for analytics and BI are daily tasks in this profession. It's a field that's always pushing the envelope, where accuracy and planning are key, and where your work has a direct impact on the power to make decisions and get strategic insights. Being a Cloud Data Engineer is an exciting and fulfilling adventure for those who are interested in a profession that combines technical expertise with data-driven strategy, and who thrive in a complex and influential work environment.

- **Cloud Data Engineer Work Environment**

  Tech businesses, financial institutions, and other data-centric organisations are common places for Cloud Data Engineers to work, where the atmosphere is usually lively and collaborative. They often find themselves in collaborative or open-plan workplaces,

places that promote creativity and new ideas. A large number of Cloud Data Engineers are now able to operate remotely because to advancements in cloud computing. This gives them the freedom to handle data systems hosted in the cloud, communicate with teams located anywhere in the world, and more.

- **Cloud Data Engineer Working Conditions**

  Intense times of concentration are common for Cloud Data Engineers, who often work full-time. This is particularly true while implementing new data solutions or fixing data system issues. Cloud platform interfaces, code authoring, and collaboration with other engineers and analysts occupy a large portion of their time. Being able to quickly adjust to new changes in cloud technology and data processing tools is an essential skill for this position. The importance of data to company operations makes for demanding but ultimately rewarding work for Cloud Data Engineers, who are essential in facilitating data-driven decision-making.

How Hard is it to be a Cloud Data Engineer?

The technical expertise of the person, the intricacy of the data systems, and the unique needs of the organisation all play a part in how challenging the work of a Cloud Data Engineer may be. Data engineers working in the cloud need a solid background in computer science, good programming skills, and extensive knowledge of database systems and cloud services. Their communication skills and knowledge of data management's technical components must be well-balanced.

Further, due to the rapid development of cloud services, Cloud Data Engineers need to be avid learners who keep up with the newest innovations in data processing and cloud computing. Though daunting, the opportunities to find answers to pressing issues, develop ground-breaking solutions, and see their efforts rewarded as companies gain a competitive edge via data leveraging are worth the difficulties. Individuals with a love for data's revolutionary potential, an eye for detail, and strong technical skills will thrive in this field.

Is a Cloud Data Engineer a Good Career Path?

An exciting and lucrative career path awaits you in cloud data engineering. The opportunity to pioneer cloud data management presents itself, an area vital to the smart and efficient operation of contemporary businesses. Talented Cloud Data Engineers are in high demand since companies big and small are relying more and more on data solutions in the cloud to power their operations and new ideas.

There is a healthy employment market, plenty of opportunity for growth, and competitive compensation for Cloud Data Engineers, according to industry trends. This is a safe and promising career path because of the role's importance to company strategy and the continuous expansion of cloud services. A job as a Cloud Data Engineer, which provides intellectual stimulation and plenty of room for advancement, is more important than ever before due to the ever-growing amount and significance of data.

7.4 The Future of Data Engineering

Modern data systems rely on data engineering to fuel many capabilities, such as machine learning and business insights. In the years to come, data engineering will be significantly influenced by emerging trends and technology.

1. The Rise of Real-time Data Processing

Overview: It used to be common practice to gather, process, and analyse data in chunks at predetermined intervals; this approach is known as batch orientation in data processing. Nevertheless, companies are clamouring for real-time analytics to help them react to market shifts nearly instantly and make speedier choices.

Technological Advancements:

- **Stream Processing Frameworks:** For processing data in real-time, leading technologies include Apache Kafka, Apache Flink, and Apache Pulsar. With the help of these frameworks, organisations may respond on new data as it comes by continuously ingesting, processing, and analysing data streams.

- **Serverless Architectures:** Developing scalable, real-time data apps without having to manage server infrastructure is becoming simpler with serverless computing platforms like AWS Lambda and Google Cloud Functions.

Impact: Efficiency in operations, satisfaction of customers, and the capacity to make proactive decisions are all positively impacted by real-time data processing and analysis. In sectors where real-time data is essential, such as healthcare, e-commerce, and banking, this change is revolutionary.

2. The Emergence of Data Mesh Architectures

Overview: Conventional data architecture often employs a centralised approach, with data housed in a single data warehouse or lake and overseen by a specialised staff. This approach has the potential to become a stumbling block as data volumes and organisations expand.

Concept of Data Mesh: A new architectural paradigm called Data Mesh is gaining traction; it decentralises data management and views data as a product. The idea is to have several groups or departments handle and provide their own data products, with an emphasis on domain-oriented data ownership.

Technological Innovations:

- **Data Products:** In the same way as software development teams oversee the final product, data product teams are liable for its availability, quality, and governance.

- **Self-Serve Data Infrastructure:** Teams will be able to construct, administer, and access data products with little reliance on centralised teams as a result of newly created tools and platforms that facilitate self-service data infrastructure.

Impact: Organisations can better handle massive amounts of data and adapt to evolving business demands with the help of data mesh designs, which encourage scalability and agility. In terms of data quality and governance, it encourages a culture of responsibility and ownership.

3. Advances in Data Privacy and Security

Overview: Strong data security and regulatory compliance are of utmost importance in light of the ever-increasing data breaches and privacy concerns. In response to these difficulties, new approaches and methods are developing.

Technological Innovations:

- **Data Encryption and Masking:** Modern developments in encryption, such as homomorphic encryption, allow for private data processing with no compromise in security. Information security may be enhanced when organisations use data masking methods to safeguard sensitive data while it is being processed and analysed.

- **Privacy-Enhancing Computation:** Data analysis may be conducted without revealing raw data using technologies such as federated learning and secure multi-party computing, which enhances privacy and security.

- **Regulatory Compliance Tools:** To guarantee that data handling methods are in line with legal standards, automated solutions are becoming more accessible to assist organisations in complying with data protection rules like GDPR, CCPA, and HIPAA.

Impact: Organisations may reduce the likelihood of data breaches, gain the confidence of their partners and consumers, and successfully traverse complicated regulatory environments by implementing enhanced data privacy and security safeguards.

4. Integration of AI and Machine Learning

Overview: To provide more sophisticated data analysis and automation, data engineering is increasingly reliant on AI and ML.

Technological Advancements:

- **Automated Data Engineering:** Tools powered by AI are automating data purification, transformation, and pipeline management, among other data engineering tasks. The efficiency and accuracy of these technologies are enhanced by their usage of ML algorithms, which detect patterns and abnormalities.

- **ML Ops:** The goal of creating ML Ops frameworks is to make it easier to deploy and maintain ML models in production settings. Model scaling, update, and performance monitoring tools are part of this.

Impact: Data engineers may increase operational efficiency, build more complex data-driven applications, and get more useful insights from data by incorporating AI and ML into their workflows.

5. Evolution of Data Storage and Management Solutions

Overview: The ever-increasing variety, velocity, and amount of data necessitates ever-evolving data storage and management technology.

Technological Innovations:

- **Cloud-Native Data Warehouses:** Organisations can handle massive datasets efficiently and affordably using platforms like Snowflake, Google BigQuery, and Amazon Redshift Spectrum, which provide scalable, cloud-native data storage and analysis options.

- **Data Lakehouses:** The term "data lakehouse" refers to a hybrid platform that combines features of data lakes and data warehouses to accommodate both structured and unstructured data. There will be less need to transfer and convert data, and the data architecture will be simplified, using this method.

Impact: Efficient data operations and better data leveraging are made possible with the help of modern storage and management systems, which also increase data accessibility, scalability, and performance.

6. The Growth of Data Observability

Overview: The capacity to keep tabs on and comprehend data pipelines is what we mean when we talk about data observability. This makes sure that data is correct, dependable, and accessible for analysis.

Technological Innovations:

- **Observability Platforms:** Organisations may find and fix problems with data quality, lineage, and performance with the use of tools like Databand.io and Monte Carlo, which provide thorough diagnostics and monitoring for data pipelines.

- **Integrated Analytics:** Analytics tools are becoming more integrated with observability systems, allowing for real-time insights into the performance and operations of data pipelines.

Impact: Enhanced data observability helps organisations maintain data quality and operational efficiency by improving the dependability and trustworthiness of data. It also makes fixing data problems and troubleshooting much quicker.

Multiple Choice Questions (MCQs)

1. **Which of the following is a common approach to integrating machine learning with data engineering pipelines?**

    a. Using machine learning algorithms to preprocess data

    b. Implementing model training and prediction within data pipelines

    c. Storing data in raw format without transformation

    d. Using machine learning for data encryption

2. **Which of the following is a framework commonly used for processing large datasets in a distributed environment?**

    a. Apache Spark

    b. SQL Server

    c. MongoDB

    d. PostgreSQL

3. **Which of the following is a key benefit of using cloud platforms for data engineering?**

    a. Reduced security risks due to centralized management

    b. Increased data latency

    c. Scalability and flexibility in processing large volumes of data

    d. Limited access to advanced tools and services

4. **Which of the following trends is expected to shape the future of data engineering?**

    a. Increased reliance on traditional on-premise infrastructure

    b. Automation of data pipelines through AI and machine learning

    c. Decrease in the use of cloud technologies

    d. Limiting the use of big data frameworks

5. **What is the role of data engineers in integrating machine learning into data pipelines?**

    a. Designing machine learning algorithms

    b. Creating features, data pre-processing, and ensuring model deployment pipelines work

    c. Performing model training and evaluation

    d. Running machine learning experiments for feature selection

6. **Which of the following is a key feature of Hadoop in big data processing?**

    a. Real-time data streaming

    b. Distributed storage and processing of large datasets

    c. In-memory processing

    d. Single-node processing

7. **Which cloud service model is commonly used for data engineering workloads where users manage applications but not the underlying infrastructure?**

    a. Software as a Service (SaaS)

    b. Platform as a Service (PaaS)

    c. Infrastructure as a Service (IaaS)

    d. Function as a Service (FaaS)

8. **Which of the following is likely to become more prominent in the future of data engineering?**

    a. Manual data entry into databases

    b. Increased reliance on batch processing for all data types

    c. Real-time analytics and decision-making using streaming data

    d. Decreased need for data cleaning and preprocessing

9. **Which of the following tools is commonly used to automate the integration of machine learning models into data pipelines?**

   a. Apache Kafka

   b. Apache Airflow

   c. Tableau

   d. MySQL

10. **Which of the following technologies is specifically designed for distributed data storage and processing in a big data environment?**

    a. Apache Hadoop

    b. Microsoft Excel

    c. MongoDB

    d. SQLite

Answer

1	2	3	4	5	6	7	8	9	10
B	A	C	B	B	B	B	C	B	A

BIBLIOGRAPHY

Abdulzahra, S. A., & Al-Qurabat, A. K. M. (2023). Data Aggregation Mechanisms in Wireless Sensor Networks of IoT: A Survey. *International Journal of Computing and Digital Systems, 13*(1), 1–15. https://doi.org/10.12785/ijcds/130101

acceldata. (2024). *What is a Data Quality Framework?* Acceldata.Io.

Ahmed, S. (2024a). *Essential Principles for Effective Data Visualization.* https://medium.com/thedeephub/essential-principles-for-effective-data-visualization-a13f05d22c39

Ahmed, S. (2024b). *Essential Principles for Effective Data Visualization.* Medium.

Alagar. (2023). *The Evolving Role of Data Engineers in Modern Businesses.* International Association of Business Analytics Certification.

AnalyticsPartners. (2023). *Data Quality; The 3 Keys To Developing A Strategy You Can Really Trust.* AnalyticsPartners.

Briney, K. (2015). Data Management for Researchers: Organize, maintain and share your data for research success. *Journal of EScience Librarianship, 4*(2), 1–4. https://doi.org/10.7191/jeslib.2015.1088

Caserta, R. K. & J. (2004). The Data Warehouse ETL Toolkit. In *Timely Practical Reliable.* https://nibmehub.com/opac-service/pdf/read/The Data Warehouse ETL Toolkit _ Practical Techniques for Extracting-Cleaning-.pdf

Chandra, B. R. (2023). *10 data engineering terms that you need to know.* Medium.

Chen Cuello. (2024). *Data Integration: Techniques and Strategies*. Rivery.

Divyansh Sharma. (2024). *Real-Time Processing Explained: Benefits, Use Cases & Tools*. HEvo.

Edwin Sanchez. (2022). *ELT vs ETL: Main Differences Between ETL and ELT (Full Comparison)*. Skyvia.

Geeksforgeeks. (2024a). *Data Acquisition System*. Geeksforgeeks.Org.

Geeksforgeeks. (2024b). *Overview of Data Cleaning*. Geeksforgeeks.

Getdbt. (2023). *Version control with Git*. Getdbt.Com.

Hattatoglu, F. (2023). *ETL Process on Data Science*. Medium.

IBM. (2023). *What is ETL (extract, transform, load)?* IBM.

Inmon, W. H. (2002). *Building the Data Warehouse* (Third Edit). Timely, Practical, Reliable. https://fit.hcmute.edu.vn/Resources/Docs/ SubDomain/fit/ThayTuan/DataWH/Bulding the Data Warehouse 4 Edition.pdf

Justin Ellingwood. (2023). *How to create and delete databases and tables in PostgreSQL*. Prisma's Data Guide.

Kleppmann, M. (2017). Designing Data-Intensive Applications. In *O'Reilly Media, Inc.* O'Reilly Media. http://oreilly.com/catalog/ errata.csp?isbn=9781449373320

Krishnan, K. (2013). *Data Warehousing in the Age of Big Data*. Elsevier. https://doi.org/10.1016/C2012-0-02737-8

Kumar, A. (2024). *23 Best Data Visualization Tools You Can't Miss!* Simplilearn.Com.

Limeup. (2024). *Batch Processing Definition*. https://limeup.io/glossary/ batch-processing/

Martin, M. (2024). *The 10 best cloud storage apps in 2024*. Zapier.

N, D. (2022). *The Data Engineering Lifecycle*. Medium.

Nadav Roiter. (2024). *Data pipeline architecture for businesses explained*. Bright Data.

Parikh, K. (2023). *Cloud Data Engineering Tools for Scalable Data Processing*. Tenup. https://www.tenupsoft.com/blog/benefits-of-cloud-data-engineering-tools-for-scalable-data-processing.html

Qlik. (2023). *Data Modeling*. Qlik.Com.

Ralph Kimball & Margy Ross. (2013). *The Data Warehouse Toolkit* (Third Edit). Kimball Group. https://github.com/letthedataconfess/Data-Engineering-Books/blob/main/Book-5Kimball_The-Data-Warehouse-Toolkit-3rd-Edition-5.pdf

Ravindra Singh. (2022). *What an effective data ingestion framework looks like*. Datavid.

Reis, J., & Housley, M. (2022). *Fundamentals of data engineering*. O'Reilly. https://soclibrary.futa.edu.ng/books/Fundamentals of Data Engineering (Reis, JoeHousley, Matt) (Z-Library).pdf

Ritu John. (2024). *Data Extraction Techniques, Methods, and Tools*. Docsumo.

Sealpath. (2023). *A Deep Dive into Encryption Types*. Sealpath.

Shukla, R., Chakraborty, A., & Joshi, P. K. (2017). Vulnerability of agro-ecological zones in India under the earth system climate model scenarios. *Mitigation and Adaptation Strategies for Global Change*. https://doi.org/10.1007/s11027-015-9677-5

Simplilearn. (2024). *Top Big Data Technologies You Must Know in 2024*. Simplilearn.

Timespro. (2024). *Data Cleaning and Data Preprocessing: Importance and Techniques*. Timespro.

Vyas, K. (2023). *Top 15 Big Data Technologies You Need to Know*. Datamation. https://www.datamation.com/big-data/big-data-technologies/

ABOUT THE AUTHOR

Srinivas Murri is a data engineering expert with over 18 years of experience in designing and implementing innovative data solutions across a wide range of industries, including retail, media, telecom, and finance. Having worked in North America, Europe, and Asia, he brings a global perspective to the challenges and opportunities in the ever-evolving field of data engineering.

Throughout his career, Srinivas has specialized in areas such as data governance, cloud platforms, real-time data systems, and scalable data architectures. He has led transformative projects that help organizations harness the full potential of their data, improve operational efficiency, and ensure compliance with global regulatory standards. Known for his ability to merge technical depth with business strategy, he has empowered organizations to make data-driven decisions at scale.

Beyond his technical expertise, Srinivas is passionate about mentoring the next generation of data engineers and sharing his knowledge with others. He thrives on helping professionals navigate the complexities of data engineering, offering guidance on everything from cloud-based data architectures to machine learning integrations. His enthusiasm for continuous learning and curiosity about emerging technologies fuel his work and writing.

This book reflects his commitment to advancing the field of data engineering. Through it, Srinivas aims to inspire and equip others with the tools and insights needed to build the next generation of data-driven solutions.